THE MARVELS OF ANIMAL BEHAVIOR

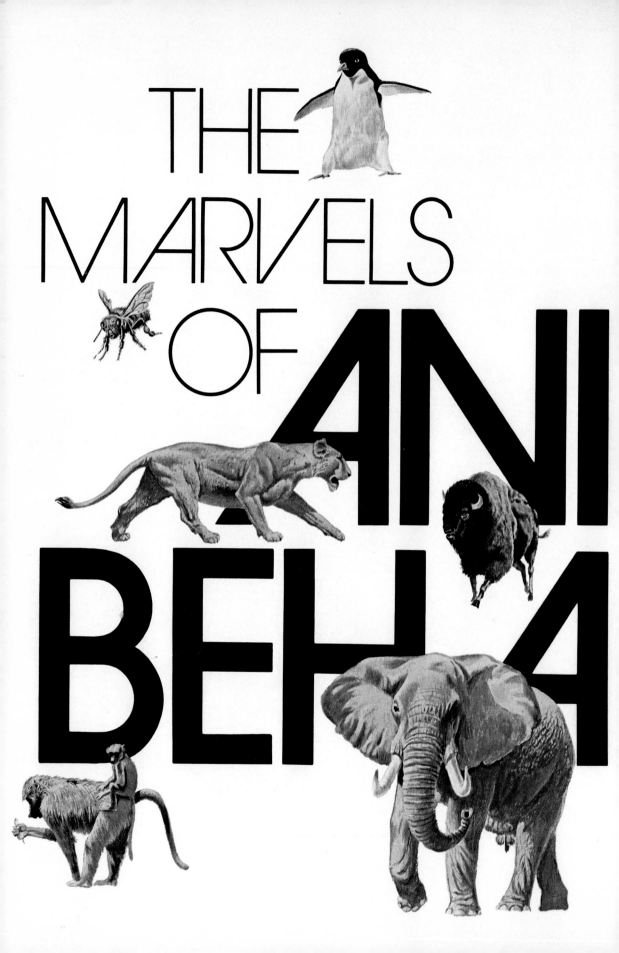

THE MARVELS OF ANIMAL BEHA

MAMMAL BEHAVIOR

NATIONAL
GEOGRAPHIC
SOCIETY

THE MARVELS OF ANIMAL BEHAVIOR

PUBLISHED BY THE NATIONAL GEOGRAPHIC SOCIETY
MELVIN M. PAYNE, President
MELVILLE BELL GROSVENOR, Editor-in-Chief
GILBERT M. GROSVENOR, Editor
FRANC SHOR, Executive Editor for this series

Editorial Consultant
PETER R. MARLER, *Rockefeller University*

Chapters by Dr. Marler and

RICHARD D. ALEXANDER, *Museum of Zoology,
University of Michigan*

JACK W. BRADBURY, *Department of Biology,
University of California at San Diego*

LEONARD CARMICHAEL, *National Geographic Society*

JOHN H. CROOK, *Department of Psychology,
University of Bristol, England*

IRVEN DeVORE, *Department of Social Relations,
Harvard University*

JOHN F. EISENBERG, *National Zoological Park,
Smithsonian Institution*

STEPHEN T. EMLEN, *Division of Biological Sciences,
Cornell University*

RICHARD D. ESTES, *Department of Zoology,
San Diego State University*

DIAN FOSSEY, *doctoral scholar in animal behavior,
University of Cambridge, England*

CARYL P. HASKINS, *National Geographic Society
trustee, Smithsonian Institution regent*

M. PHILIP KAHL, *research scientist,
International Council for Bird Preservation*

HANS KRUUK, *Department of Zoology,
University of Oxford, England*

DALE F. LOTT, *Wildlife and Fisheries Biology,
University of California, Davis*

AUBREY MANNING, *Department of Zoology,
University of Edinburgh, Scotland*

WILLIAM A. MASON, *Department of Psychology,
University of California, Davis*

GORDON H. ORIANS, *Department of Zoology,
University of Washington*

ROGER PAYNE, *research zoologist,
New York Zoological Society*

RICHARD L. PENNEY, *research zoologist, author*

ARTHUR J. RIOPELLE, *Department of Psychology,
Louisiana State University*

GEORGE B. SCHALLER, *research zoologist,
New York Zoological Society*

WOLFGANG WICKLER, *Max Planck Institute
for Behavioral Physiology, West Germany*

A VOLUME IN THE
NATURAL SCIENCE LIBRARY
PREPARED BY
NATIONAL GEOGRAPHIC BOOK SERVICE
MERLE SEVERY, *Chief*

Staff for this Book

THOMAS B. ALLEN
Editor

SEYMOUR L. FISHBEIN
Associate Editor

CHARLES O. HYMAN
Art Director

ANNE DIRKES KOBOR
Illustrations Editor

**ROSS BENNETT,
JULES B. BILLARD,
EDWARD LANOUETTE,
DAVID F. ROBINSON,
VERLA LEE SMITH**
Editor-Writers

WILHELM R. SAAKE
Production Manager

**JOHN R. METCALFE,
WILLIAM W. SMITH,
JAMES R. WHITNEY**
Engraving and Printing

**JAN NAGEL CLARKSON,
DIANE S. MARTON,
HARRIET H. WATKINS**
Editorial Research

CONNIE BROWN, *Design*

KAREN EDWARDS, *Production*

**BARBARA G. STEWART,
DAVID H. McCUEN,** *Illustrations*

**ESME BENJAMIN,
CONSTANCE MOSHER,** *Assistants*

**JOHN D. GARST, VIRGINIA L. BAZA,
BETTY CLONINGER, NANCY SCHWEICKART,
MONICA T. WOODBRIDGE**
Geographic Art

WERNER JANNEY, *Style*

TONI WARNER, MARTHA K. HIGHTOWER
Index

419 illustrations, 390 in full color

*Locked in combat, monitor lizards fight for
territory—a key to survival in the animal kingdom.
Defending his ground in Western Australia,
the territory owner pits his experience and his
natural endowments—size and strength—against
a slimmer male invader. The landholder won this
match by repeatedly tripping his opponent.
These sand, or ground, goannas vary in color and
pattern and may grow to nearly five feet in length.*
ARMAND D. MASTERS

CONTENTS

MAN AND ANIMAL

Leonard Carmichael

A NEW UNDERSTANDING

Man is a talking animal. Man's brain has qualitatively little to distinguish it from the brains of his close cousins, the gorilla and the chimpanzee. But the human brain is larger, and this quantitative difference somehow sets man apart. His brain alone allows the development of complex language and the ability to read, write, evolve mathematics, and invent machines and social systems. The capacity to recognize, establish, and obey religious, legal, and esthetic codes also depends on this unique human computer. Apes can make tools and communicate in simple ways. Only people write poems.

Men have not always been glad to admit it, but from earliest times they have had a deep and almost certainly inborn interest in other animals. It is still evident in modern man. There are more than 500,000 pet dogs in New York City alone. Zoos attract more people than ever before, as do African game-watching safaris.

This book reflects this ancient interest. It is written by experts in the modern study of animal behavior, and especially in the new science called ethology. In some ways ethology resembles old-fashioned natural history, for both subjects concern animals as they live their lives in the wild. But ethology is as new as the Atomic Age; its use of scientific procedures and devices produces a picture of the animal world far more accurate than that shown in the writings of Victorian naturalists.

The authors of this book set forth this new science's broad themes in chapters that reveal how animal societies work, how animals communicate, how they fit into their environments, how they court, mate, and raise their young, and how they learn. Ranging through communities of bee and elephant, ant and whale, we perceive on every page a unity of behavioral patterns. As we read on, we recognize many traits that we share with our nonhuman companions in the living world: social bonds, maternal care, aggression, submission, and ownership of territory; even, alas, deception and greed.

Though our scientific guides are modern, our trek is not. We are on a newly blazed path in an ageless search for an understanding of how our fellow animals live and act. The quest is as old as the fossilized remains of very early man. The fossils tell us that these ancestors of ours in Africa lived at least partly on meat, cracking open bones to get at the marrow of scavenged animals or those they killed. Human beings have always been able to live entirely on vegetable foods, and millions do so now. But in all ages and in most civilizations, when mammals, birds, fish, or even insects have been available, man has generally been in part carnivorous. So man has always been interested in other animals for one very basic reason: He wants to eat them.

Some scientists believe that man may have been a scavenger at first, eating only the

Messmates on a High Sierra pack trip share food and friendship, affirming a timeless bond.
LOWELL GEORGIA

*Toiling together, man and his domesticated animals span time
from Stone Age painting to 20th-century sheep range. The fresco,
some 4,000 to 6,000 years old, adorns a sandstone grotto near
the center of the Sahara. Evoking an era when the area abounded
with life, the tableau portrays herdsmen driving their cattle home
from the fields; two hitchhike on an ox's back. A dog accompanies
them, perhaps helping—as do sheep dogs working a flock
in Nevada (right). The shepherd's best friend responds to whistles
or hand signals but can also round up strays on its own.*

DOCUMENT MISSION HENRI LHOTE. RIGHT: WILLIAM BELKNAP, JR.

remains of animals killed and abandoned by more powerful predators, such as the big cats.
Eventually man himself became a hunter, and as he developed weapons he became what
he still is—the greatest of all predators.

Anthropologists often speak of the earliest period of human culture as the hunting and
gathering stage. During this time, before agriculture began, men stalked, killed, and ate
wild animals, and gathered wild berries, grains, and roots for food. Of necessity, those
earliest of hunters, like today's Nimrods, had to know the ways of their quarry. Successful
hunters and fishermen are always animal behaviorists.

But man being man, there was more to his life than the filling of his stomach. As he
gained skill in the chase, he inevitably thought much about the lives of the animals he
pursued. Thus, in time, some artistic hunters began to paint representations of animals on
the walls of their caves. Others modeled animal likenesses from clay or carved figurines
of bone or stone. There is evidence that often metal and clay were fashioned into animal
and human figures before these materials were used to make weapons or pots.

Primitive man's knowledge of animals is best illustrated by the fact that, with a few
possible exceptions, all the domesticated animals known today were developed by his
patience and skill. We can trace to prehistoric times horses, asses, camels, llamas, goats,
sheep, swine, chickens, ducks, geese, turkeys, rabbits, guinea pigs, cats, and dogs.

Our varied modern dogs are all descended from the wolf. Obviously, the domestication

of such animals as the dog required that early man select the most useful or tractable young as a breeding stock. As such selection continued in the dog and other species for countless generations, animals truly adapted to man were produced. They became pack carriers, givers of milk, suppliers of meat and blood, fur and hides. But man did not see all animals as providers for his physical needs. He wanted some animals to revere.

Hundreds of animal drawings adorn the walls of that gallery of prehistoric art, the Lascaux Cave of France. One that particularly intrigues anthropologists shows a bison and a man. The bison is realistic; the man has the head of a bird. Near his right hand is a stick bearing the figure of a bird. Thus some 15,000 years ago did a Stone Age artist bequeath to us perhaps the earliest known evidence of totemism, a belief in a mystical kinship between man and animal.

The belief lived on, far beyond the Stone Age. Many tribes, especially in pre-Columbian America and in pre-missionary Australia and Melanesia, revered certain animals because of the special blood relationship they were thought to have with members of the tribe. The social unity of the tribe was based on the ghostly presence of the animal ancestor of each member. Its likeness was tattooed on tribal members; its head appeared on masks and weapons — or loomed on what we call totem poles. Taboos often prevented the killing or eating of the totemic animal. For some people today eating horseflesh is taboo.

All that we have learned about prehistoric and primitive man thus suggests that his dreams and his earliest approach to art emphasized the animals he knew. The idea that modern man has certain "atavistic interests" is not scientifically popular today. But some of us still trace our modern interest in animals—and love of them—back to deep-seated spiritual and religious attitudes similar to those held by our early ancestors. Even such great 20th-century psychiatric writers as Freud and Jung have not passed by this idea without at least making an intellectual genuflection in its direction.

Animals have persistently appeared in the art of vastly different cultures. Look at the similar man-and-animal motifs in the paintings and sculpture of ancient Egypt and in the gargoyles and heraldry of medieval Europe. Even in the modern art created in developing or advanced countries we can see dreamlike representations of the birds and beasts that recall Victor Hugo's word picture—"the visible phantoms of our souls."

As man moved from primitive totemism to the paths of religion, animals remained with him. Adherents of the doctrine of metempsychosis, the transmigration of souls, believe that the animating principle passes at death from one body to another, and often during

"...with fervent hearts we praise you," reads an Indian poem in honor of serpents. Women in Shirala

a series of reincarnations the human ghost exists in various animals in turn. Some Hindus still worship snakes and bulls as incarnations of divine beings.

The Bible, which mentions some 100 kinds of animals, is a veritable treasure-house of references to man's relationship with "the fish of the sea . . . the fowl of the air, and . . . every creeping thing that creepeth upon the earth." Biblical commentators often showed insight as animal behaviorists. Jeremiah (8:7) notes the migration of birds: "Yea, the stork in the heaven knoweth her appointed times." The description of "behemoth" in Job (40: 15-24) has been hailed by the distinguished zoologist Sylvia K. Sikes as an "exquisite summary of the elephant and its natural history." Biblical laws decreeing kindness to animals are the cultural source of some of our own humane laws.

In classical antiquity some speculative philosophers gradually began to adopt the methods of thought and inquiry that we now call science. As this great intellectual revolution developed, animals came more and more to be not just objects of mystery and worship and food but also subjects worthy of careful study. Aristotle, who wrote extensively on biology, knew enough about bee behavior to note that hives had "rulers," workers, and

venerate a cobra by prostrating themselves within striking range—the length of the raised part of its body.
NARESH AND RAJESH BEDI

drones. He described animals in terms of their differences and similarities, both in structure and in function. Aristotle's ideas continued to influence academic thought about animals for centuries. Indeed, his classification of the living world was not greatly improved upon until the work of such men as Linnaeus in the 18th century gradually corrected his errors. Aristotle believed that some birds spent the winter beneath the sea, but he also realized that changes in a bird's physiology related to changes in the seasons—a phenomenon known to Homer and which scientists still study today.

As early as the ninth century of our era, Arab scholars introduced Aristotle to Islam, and able scientists in the Moslem world taught his works on animal life. During the Middle Ages in Europe, Jewish scholars studied and developed the Aristotelian view of the animate world, as did St. Thomas Aquinas and other Scholastic philosophers. In a bestiary, or book of beasts, of that same time we can read that when certain fish "sense some kind of danger" threatening their young "they are said to open their mouths and thus hold the little ones." A medieval myth? No, a marvel of animal behavior that a modern ethologist describes in this book—and which you can see on page 311.

The founding of the science of ethology is often credited to Niko Tinbergen and Konrad Lorenz [winners, with Karl von Frisch, of the Nobel Prize in 1973]. Though they are objective scientists, they love animals. Dr. Tinbergen's career traces back to a summer day when, after watching hundreds of wasps return to their "town," he wondered how each managed to find its own burrow. Dr. Lorenz has never lost his similar sense of wonder— nor his belief that "our fellow creatures can tell us the most beautiful stories."

Wonder, curiosity, fascination—we find these qualities also among early workers in animal behavior. D. A. Spalding in 1873 reported that as soon as chicks were able to walk they would pursue any moving object, with "no more disposition to follow a hen than to follow a duck, or a human being." Decades later, this behavior would be labeled "imprinting." A German, Oskar Heinroth, and two Americans, C. O. Whitman and Wallace Craig, early in this century demonstrated that early imprinting influences adult behavior. Dr. Craig, curious about the effects of isolation, reared male doves that, in maturity, initially ignored a female dove, choosing rather to court the experimenter's hand!

Today the work of ethologists clearly shows that man is united with all animals. From ethology we get extremely useful ideas about the unlearned behavior of human infants and the psychological attitudes of adult men and women. Irenäus Eibl-Eibesfeldt, for example, sees resemblances between the "ritualized attack movement" of a man stamping his foot in anger and the foot-thumping of a badger or squirrel threatening an opponent.

Animals, now psychological teachers of man, long have been his biological teachers. In the days of Imperial Rome, Galen, the court physician to Marcus Aurelius, made himself a student of animals. He dissected monkeys and many other species, using the information he gained to advance his knowledge of biology and medicine. This understanding led him

Pioneers in the science of animal behavior: I. P. Pavlov, the Russian Nobel laureate, measures a dog's saliva flow in experiments proving animals could be conditioned to respond to neutral stimuli. Off a Florida key, Konrad Lorenz, Nobel laureate and a founder of modern ethology, collects sea creatures and transforms a jar into a lab. The late Dr. Carmichael, who studied prenatal development of behavior in mammals, observes chimpanzee research at Gombe Stream National Park in Tanzania.

15

HUGO VAN LAWICK. FAR LEFT: BATES LITTLEHALES, NATIONAL GEOGRAPHIC PHOTOGRAPHER. UPPER: SOVFOTO.

to the realization that arteries carried blood, not air. He also made discoveries about the brain and the nervous system.

Throughout the history of medicine, the study of animals has continued to aid scientists. By work with animals, William Harvey demonstrated the circulation of the blood and Sir Charles Bell showed how the sensory function of the nervous system was distinct from its function in activating muscles. In our day, many animals are bred for laboratory studies. The Jackson Laboratory in Bar Harbor, Maine, for example, each year raises more than 2½ million mice of various inbred and mutant strains. They are used all over the world for research on the cause and cure of cancer and such other diseases as muscular dystrophy and diabetes. Lovers of animals should also remember that veterinarians, who can now do so much to cure diseases and keep pets healthy, depend on knowledge gained in past experimental studies of animals by anatomists, physiologists, and pharmacologists.

Much modern knowledge of human neurology and psychiatry can be traced back to experimental work with animals. The great Russian scientist I. P. Pavlov won a Nobel Prize for his discoveries about the physiology of digestion; much of this research was done on dogs. Later, still using dogs, he discovered what we now call conditioned reflexes. He gave us insights into a method of human learning and how the human brain works. His studies have influenced modern care of the mentally ill. Knowledge of conditioned reflexes has also helped us to develop an understanding of the way in which some school subjects are most effectively taught.

In the present-day work of B. F. Skinner, an American specialist in learning and habit formation, we can once again see how a knowledge of animal behavior can lead to an understanding of human behavior. Dr. Skinner, by studying the acquisition of new skills by pigeons, rats, and other animals, illumined the roles that rewards and punishments — "positive" and "negative" reinforcements — play in learning. Other modern research, especially that by comparative psychologists, has increased our understanding of the mental lives of animals. Today we know much about how animals perceive, how they react emotionally, how they communicate.

Now, through ethology, scientists turn from the laboratory to the wild, seeking to explain how animals organize their societies and adapt to their own natural surroundings. In this book, we meet members of the new generation of behaviorists and journey with them to see animals where they live. We will learn about animals in their own world — and perhaps learn more than a little about how we live in ours.

For as we read this book about wild animals we will do well to bear in mind that we are of a mammalian species called *Homo sapiens*. That name was first applied to human beings by the great biologist and classifier Linnaeus. A little boastfully, it suggests that man is the wise primate. But now it seems if we are to be really wise we must amend Alexander Pope's observation: "The proper study of mankind is man." If Pope were writing today, he would be wise to add a postscript: ". . . and nonhuman animals."

In the dawning of a new African day, a browser dines in solitude on the Serengeti Plain; the giraffe's height makes it visible to others of its kind up to a mile away. On this seemingly limitless sea of grass dwell such gentle giants, packs and prides of hunters, great herds of the hunted. Studying the ways of these animal societies, zoologists gain new knowledge of man's fellow creatures.

GEORGE B. SCHALLER

Dr. Leonard Carmichael, a Vice President
of the National Geographic Society, died
on September 16, 1973. In helping to guide this book
to its first printing in 1972, he continued his
lifelong effort to increase our knowledge of ourselves.

THE DRIVE TO SURVIVE

Peter R. Marler

Charles Darwin, in *The Origin of Species*, remarked that if man could impose his standard of beauty on domestic fowl by deliberate breeding, then female birds could certainly impose their standard "by selecting, during thousands of generations, the most melodious or beautiful males." In the ornamental poultry of his time—the diminutive Bantam of "elegant carriage," the blue-wattled, frizzy-feathered Silkie—Darwin saw man's short-term version of a timeless natural process.

As marketing demands or fanciers' whims veer this way and that, so animal breeders select their stock accordingly. They cull individuals with undesired traits, and keep for breeding the brothers and sisters that have even a slight edge toward desired traits. Repeat this process with each generation and the result is constant change in the structure and behavior of domestic animals. Numerous varieties can derive from the same stock. The American Kennel Club recognizes 120 breeds of dog, only some of the many descendants of the wolf produced by 8,000 to 10,000 years of domestication.

Like farm breeds and pets, wild animals also change, the forces of nature substituting for the breeder's eye. Success in such matters as evading a predator or overcoming environmental hazards favors some individuals and their offspring. Darwin called this process, "by which each slight variation, if useful, is preserved," natural selection—the principle that his genius recognized as the root of biological diversity.

The rate of change is perhaps 300,000 times greater in the breeding of animals by man than it is in evolution, no doubt because natural selection is less intense. And nature is usually selecting for a balanced set of characteristics that makes for success in a variety of circumstances. It is a slow and intricate business.

We can all think of dog breeds that in some ways excel the wolf. The lean-limbed greyhound might keep pace with stampeding caribou more readily. But the greyhound could hardly survive the rigors of an Arctic winter, and, lacking muscular jaws and strong teeth, could not fall back on a diet of bones in hard times. It has been aptly pointed out that no dog is a superwolf. Each dog is a specialist that pays some price for its talents. In nature, the proper balancing of such talents is vital—and in such equations behavior counts as much as bodily structure.

Earth has more species of animals than we have yet been able to count. So far we have classified about 32,000 species of fishes, 8,900 species of birds, 6,000 mammals, 6,500 reptiles, and 2,500 amphibians. Insects are innumerable; estimates of still undescribed species run as high as 9 million. All these species evolved as a result of natural selection and each adapted to a different set of conditions, a different niche. All also belong to a network of shared heritage, which is why we have so much in common with other

Old as life, stark as tooth and claw, the endless struggle for survival shapes earth's every creature.

19

PAINTING BY RICHARD SCHLECHT

mammals. Medical science assumes that physiological studies of rats, dogs, and monkeys can be translated into human terms. Our bodies share with many other mammals the same basic design, from cells to vital organs. Every major part of the human brain, for example, is represented in the brains of cats and dogs.

The notion of physical kinship between man and animals seems readily acceptable, but not so psychological kinship. We are tempted to think that any parallels are remote and irrelevant. But I believe that animals and man have much more in common than we realize. Studies of animal societies to gain knowledge of the human condition will prove to be as relevant in the psychological realm as in the physiological. If this is not yet obvious, it is because our understanding of animals is still so incomplete.

To us, the song of a bird is a source of delight; to the birds, singing is essentially a social activity—though perhaps not lacking in pleasure for them as well. The female finds the male's song literally attractive, and she responds with signs of readiness to cooperate in breeding. A male, hearing another male's song, senses a threat and reacts as a competitor. The challenged male, after retreating to his own territory, will hurl his own challenge back. Such male counter-singing across a territorial boundary may continue for hours.

During the breeding season, when birdsong is at its height and territoriality at its most imperative, songs and calls help to sustain social bonds between mates, between parents and their young, and perhaps even between chirping competitors. If we learn how to listen, we can hear in the song of birds the point and counterpoint of survival and of all social behavior: cooperation and competition. In order to survive, all animals must keep the two themes in harmony. Finding how that harmony is achieved, exploring the laws that govern the balance between cooperation and competition—these are major tasks of ethology, the science of behavior.

An animal may do some things alone, such as finding food or avoiding predators. But a surprising number of animals choose to cope with these and many other problems of life in the presence of companions. Sometimes animals join companions only to compete with them. Even members of the same family compete, as anyone knows who has watched kittens squabbling to nurse from their mother.

To understand social behavior we must understand the universal drive to compete. And we must realize that the mutual benefits of social cooperation always have to be weighed against the drawbacks of living socially. The point of balance between the forces of competition and cooperation shifts for various species. Some animals will readily form a peaceful group, while others are always fighting. Onto the scale go the special attributes of the individual's own body and the emotional inclinations that are rooted in the species' heritage. But the delicate balance can be swayed by such external factors as the availability of food and whether an animal is hunter or hunted. From these facts of life—facts that ethologists try to discover—come the reasons why two species created from the same stock can behave as differently as the lone jackal and the gregarious hyena, the bumblebee that starts life on her own and the honeybee whose hive is the acme of sociality.

We can find sociality even among organisms once labeled "lower," such as algae and sponges. Their colonial life-style calls for individual cells to share the labors of the cellular commune. If we look back far enough, the roots of (Continued on page 30)

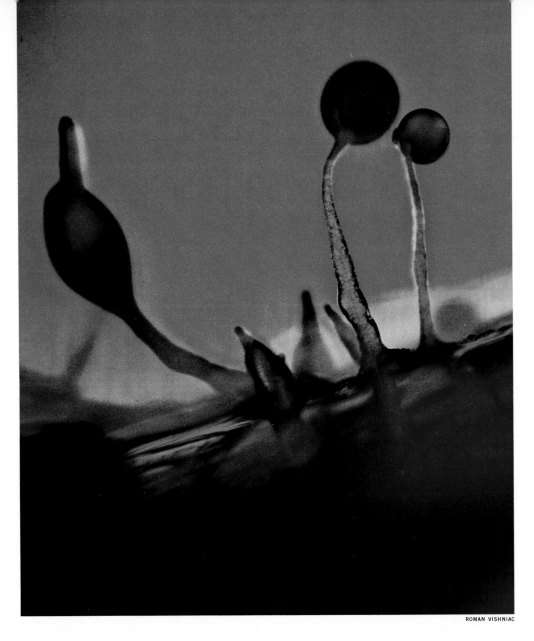

GETTING TOGETHER

Oozing across a microscope's stage like menacing blobs in a science-fiction movie, members of a strange community answer a summons to social living in a slime mold. Free-moving, single-celled, independent amoebas—each the size of a human white blood cell—feed on bacteria in the soil. When the food supply dwindles, they begin aggregating, attracted by a substance produced by some of the cells. As the amoebas stream toward a focus, they become sticky and adhere to each other, forming a cylindrical "slug." This eventually gives rise to a complicated, spore-bearing body that is a kind of fruit on a stalk (above). No individual could have accomplished the feat alone. The amoebas had to pool their resources. And, as we say about our own families, clubs, and teams, they had to stick together—the essence of social behavior in animals large and small.

21

Communes of the sea

The Portuguese man-of-war, formidable as its namesake, is a living vessel, a colony-like animal of four distinct kinds of organs called polyps, each type serving a function. One group forms tentacles that paralyze prey with a toxin as potent as cobra venom, then reel in victims. Other polyps, virtually all mouth, specialize as the man-of-war's eaters and digesters. Sexual polyps produce eggs and sperm. A fourth type creates the buoyant gas-filled float, whose crested comb is a tops'l that catches the wind.

Another type of polyp, the orange tube coral at left grows in colonies—but does not build reefs, as do most tropical corals. Mobile sea creatures often seek safety in society. For fish, it's the school; gray snappers below coalesce to flee danger.

RUSS KINNE, PHOTO RESEARCHERS. FAR LEFT: CHARLES E. LANE. UPPER: DOUGLAS FAULKNER

Social security benefits

In a circle less than an inch wide or in a woolly wall that blocks the Arctic horizon we see arrays for survival. Stinkbugs, abandoned by their mother, encircle their eggshells in a communion rite. Shoulder to shoulder, adult musk oxen—each weighing 400 to 800 pounds— form a bulwark between their young and their enemies. Bug or behemoth, animals find in society both sustenance and a shield.

Clumped, the stinkbugs intensify the repellent smell and color pattern which reminds a predator how vile they taste. Around their empty eggshells they absorb their mother's parting gift—some of her intestinal bacteria left on the egg surfaces. Each bug sucks into its gut a lifetime supply for digesting plant juices.

The musk oxen, shaggy relics of the Pleistocene, evolved the wall defense against wolves. Pushed south by ice sheets, the animals confronted a new enemy: primitive hunters whose spears and arrows mowed down the immobile walls. Slaughter exceeded the speed of adaptation. So engrained is the defensive tactic that even calves will form a circle if threatened. Musk oxen were nearly wiped out in modern times. Now, coveted for their cashmere-like wool and easily bred in captivity, they are on their way toward becoming the latest in the long line of animals domesticated by man.

ROBERT W. MADDEN. OPPOSITE: EDWARD S. ROSS

The tugs and ties of society

Wild stallions rear and fight each other. Macaques groom each other. Entwined in both acts are competition and cooperation, the dual monitors that regulate an animal society and its members.

A duel between stallions begins with a face-off that allows either to back down. Only if this nonviolent competition does not work do the horses battle. The winner gets a harem he must protect. In hazardous travel, he trails the harem, alert to enemies. If he must make a stand against them, the lead mare directs the escape.

The macaque, swooning with delight at a companion's touch, registers reaction with a signaling device used by many animals: the face. Slight changes in expression can make a greeting a threat.

Such a transformation of gestures is a performance that Darwin called the Principle of Antithesis. He noted it when a "dog suddenly discovers that the man he is approaching, is not a stranger, but his master." Instantly, the dog slinks, wags his tail, sleeks his bristled hair, and averts his gaze—all in contrast with the pose of threat.

EIJI MIYAZAWA, BLACK STAR. OPPOSITE: HOPE RYDEN

The lively arts of coexistence

Neighbors drop in from the sky—one to kill, the other to feed. Such encounters make the animal kingdom a social maze that all must master to stay alive. The Virginia opossum and her young confront a red-tailed hawk in a classic tableau of prey vs. predator. The red-billed oxpecker, perched on an impala buck's neck, snatches a tick from the ear of its host and gets a meal; it may even harvest loose hairs for its nest. The transaction, known as mutualism, is a form of symbiosis, a social contract binding many species.

Bird-and-mammal symbiosis is sometimes exotic. The African honey guide's cries lead the honey badger—and men—to a hive. When the followers open the hive, the bird feasts on beeswax.

The opossum, baring her teeth and hissing with accompanying chorus, frightened off the young hawk. Alone, she might have "played 'possum" by feigning death. Her young are themselves survivors of a struggle begun at birth. A marsupial whose newborn live in a pouch, she may bear more young than she can feed. Each needs its own teat, which it locks onto continuously for about 60 days. Losers in the scramble for feeding stations starve to death.

ALLAN ROBERTS. OPPOSITE: KARL MASLOWSKI

social behavior mingle with the origins of these many-celled organisms. In fact, even one-celled bacteria show something equivalent to sex differences.

Invertebrates, that vast multitude of animals which range in size from the microscopic one-celled protozoans to the giant 120-foot squid, may not have backbones but they do have a social life. The pioneering American investigator of animal social behavior, W. C. Allee, used invertebrates to illustrate many of his theories. He showed, for example, that many organisms survived better in large groups than in small, as long as they were not overcrowded. His book, published in 1938, introduced many people to the concept stated in its title, *Cooperation Among Animals*.

In one of his experiments he demonstrated that when brittle starfish were deprived of their normal eelgrass cover, they clumped together, "substituting each other" for the strands. "It is a far cry from such aggregations to the groupings of foreigners in a strange city that result in Little Italy, or the Mexican settlement, or a German quarter," he wrote, "and yet basically some of the factors involved are similar."

The great British philosopher Herbert Spencer, who anticipated Darwin on some principles of evolution, drew a parallel between societies as aggregations of individuals

TASI GELBERG PESANELLI, INC.

3

A monarch butterfly displays coloration that seems to make it a target for insect-hunting birds. But its wings and body still contain poison. Predators lured to this uncamouflaged prey will learn a lesson that will be distasteful—and most likely unforgettable.

1

A saga of survival, worked out by L. P. Brower of Amherst, begins with this milkweed that grows in Southeastern states. A poison within it can give a vertebrate heart failure. But the toxin carries its own antidote: half a lethal dosage induces vomiting. Shunned by grazing animals, the plant survives for another role.

2

A monarch butterfly larva chews on a milkweed leaf, ingesting Asclepias humistrata's poison. An invertebrate, the larva is not affected. The chemical, similar to the human cardiac drug digitalis, is assimilated in the larva's body. Monarch larvae dine exclusively on milkweed, only some species of which are poisonous. Larvae bearing poison may live in the same colony with larvae bearing none. All will metamorphose into monarchs of identical appearance.

and organisms as aggregations of cells. The analogy is especially tempting with termites, ants, and bees; among such social insects a whole colony could have descended from the same father and mother—much as the cells of an organism derive from the union of one egg and one sperm. Some organisms may indeed be best thought of as clusters of individuals, tethered together by a common fate. Coral reefs are no less than the accumulated skeletal remains of large colonies of tiny, united polyps. We are only beginning to understand the complexity of the Portuguese man-of-war, whose components are externally so dissimilar that they seem to be different organisms rather than organs.

We can find the benefits of cooperative living not only among kin and in a species but even beyond. Bees and flowering plants, about as distantly related as any pair of organisms you can name, engage in cooperative ventures—interactions that have become fundamental to the biology of both. Bees pollinate flowers, which yield nectar and pollen in return; this is a true symbiosis, literally a "living together."

Generally, the more that two animals differ in form and in needs, the less chance they will compete with one another. Indeed, the impact of competition has led forms to diverge from one another—and thus create new species.

4 *Enter a blue jay, dubbed "naive" by behaviorists because it has never tasted an unpalatable monarch. The jay eats, vomits, and, remembering the experience, rapidly loses its naivete. Though one or more monarchs may die to teach a jay the lesson, most survive—including tasty ones that are not poisonous.*

Now a sadder but wiser bird, the jay avoids monarchs as well as mimics that have evolved the protection of looking like monarchs. Jays in a lab retched just seeing a monarch! If jays' hearts were stilled, the ecological community would lose an educated predator species that helps stabilize a survival system based on masquerade. Even plants may mimic, fooling grazers by looking or tasting toxic. 5

Royal terns hover like a cloud over their colony—a grid of egg-laden sand nests that may be packed six

Animals in a species have so many characteristics in common that they are forced to compete with each other. Such competition may be indirect, as when one animal's possession of a food source deprives others of it. Or competition may be direct, perhaps face-to-face. Much animal behavior deals with resolving competitive conflicts—activities that we often speak of collectively as aggression. An animal may repel or subjugate another by defeating it in actual combat. But this is surprisingly uncommon. Instead, animals have evolved symbolic displays whose meanings are mutually understood. Such saber rattling serves the same function as combat, with less danger of irreparable harm.

Territorial behavior—defense of a piece of land against intruders—is the most familiar manifestation of aggressive tendencies. Sometimes the conflicts of territoriality flare not merely between members of the same species but between species. Certain western marshes provide nesting sites for three blackbird species: yellowheads, tricolors, and redwings. Ecologist Gordon H. Orians (page 169) found that each spring the redwings settle first and invariably claim more land than they can hold. When the other species arrive,

to a square yard. Neighbors squabble over their tiny territories but unite for defense against predators.

some redwings must withdraw, either because of aggressive eviction proceedings by yellowheads or because the redwings find themselves overwhelmingly outnumbered by tricolors. A short supply of a commodity needed by more than one species — in this case, marshland nesting sites — has called forth territoriality between species.

If a species lives in a place that can satisfy lifetime needs of its members, the area may become permanent territory. The ecology of some monkeys and apes thus permits them to settle, though this does not mean that territoriality is as fundamental to primate biology as, say, development of the hand. True territoriality does occur in some primates, such as Asian gibbons, that live in the upper level of the forest. Primates that spend more time on the ground and invade open areas tend to behave differently.

On the African savanna, for example, the home ranges of baboon troops overlap extensively. Part of a troop's range remains uninvaded by neighboring troops so that the residents have exclusive use of that part. Exclusion of neighbors does not require aggressive defense by baboon troops. A more subtle avoidance of each other's presence suffices. 33

Among chickens, the bill gives rights: Social order is based on a dominance hierarchy, with high-ranking hens pecking lower ones. The red dots on the chart represent the number of pecks each of 11 hens received in the setting up of a 12-hen peck order. Hen No. 2 was pecked only by No. 1; both pecked 3, and so on down the line to No. 12, pecked by all—and pecking none. After the order is fixed, fights give way to threats and bows. Top hens get priority at feed and water, dusting areas, and roost. The system, also used by cocks, eases social tensions.

Home ranges may shift somewhat from season to season; a troop may need to get to water and better grass in the dry months—or to trek to trees with particularly delectable fruits. But for other animals, the seasonal wanderings may be more drastic, and we find a complete change in the organization of aggressive behavior from one season to the next. This is what happens with many songbirds of temperate climates.

As a graduate student at the University of Cambridge in England, I studied chaffinches for several years. In spring and in summer, nearby Madingley Wood resounded with the territorial singing of males. Each occupied an area of a few acres. From the area he expelled all other males, and to it he admitted females. Once he took a mate, she would repel other females. Thus the pair had exclusive use of the territory for the breeding season, ensuring an adequate supply of food for their young. In autumn, territoriality waned, and the birds left the woods to form large flocks that wandered through the meadows and farmlands of East Anglia.

Despite the abandonment of territories, the chaffinches still quarreled at their winter feeding table—but by a different set of rules. Although aggression was still associated with space, this was not the fixed space of a territory. Each bird related to a smaller area that it, in effect, carried around. We can think of this as "personal space." If another bird penetrated it, then aggression very likely broke out and the birds separated.

In winter, birds may wander so far and wide in search of food that a territory would have to be impossibly large. So it makes sense for the defended area to shrink to a small space around the individual. It also makes sense that a bird which happens to approach too closely will provoke aggression. By repelling an intruder, a bird may defend a meal of seeds until it can eat them all. The defender then moves on.

In some species, each duel of aggression seems to be a fresh trial of strength and endurance. The circumstances determining the winner vary from encounter to encounter. Among winter flocks of chaffinches, though, I found that when males were pitted against females, the males always won, usually without a fight.

I once captured a flock to study the fighting. As soon as I had learned to identify all the individuals, I discovered that the outcome of any fight was predictable to some extent. The pattern of fighting revealed a dominance hierarchy: One animal at the top wins encounters with all others in the flock; the second in rank wins all except those with the top-ranking bird, and so on down the line until the lowest in rank loses all its fights.

We can easily appreciate how such peck-order hierarchies come about. Imagine a first-time meeting between two birds, one smaller than the other. The larger probably wins, unless he has some handicap, such as meeting on his opponent's own grounds. After a few defeats, the smaller bird will learn to submit not only to his opponent but also probably to all others that are larger. Since chaffinches, like any experienced bird-watcher, can readily identify individual birds, the memory of past encounters will influence behavior in future disputes. So the birds can eventually anticipate the outcome of each encounter, and a dominance hierarchy is the result.

Such behavior does not always depend on learning. To demonstrate this, my colleagues and I raised young female chaffinches by hand. We dyed the grayish-brown breasts of some with the ruddy color of the male's breast. Like true males in a winter flock, most dyed females dominated other females—though none had ever seen a male before!

Sometimes the separating or congregating of animals is inspired not by the desire to eat but by the need to keep from being eaten. Gardeners are all too well aware that some caterpillars, such as the green, beautifully camouflaged larvae of the cabbage looper moth, are spaced out, difficult to find—and, at least to birds, good to eat. An infestation of sawfly larvae on an ornamental tree confronts the gardener with quite a different situation. Instead of spacing out, these diners gather in dense clusters. Unlike their hidden, camouflaged colleagues, the sawfly larvae are conspicuous in behavior and appearance—and bad tasting. This contrast reflects the ways that insect species try to keep from becoming meals for birds, which consume vast numbers of larvae.

Consider the problem from the viewpoint of a hungry blue jay. As he looks for a meal, he does not search at random, but favors a strategy that has worked in the past: hunting among the leaves at the very tips of tree branches or hopping along the ground under cabbage plants, eyeing the undersides of the leaves. His persistence will depend on how rapidly he is rewarded. Even when food is choice, if he must search for a long time, he is likely to abandon that strategy for another that yields a quicker payoff. From the viewpoint of the cabbage looper caterpillars, however, their chances of surviving are obviously much better if they are hidden and far apart.

Now suppose the jay finds a group of bad-tasting sawfly larvae for the first time. Like many species that have evolved stingers, warning coloration, or chemicals noxious to birds, the sawfly larvae cluster. Their interest will best be served if a blue jay can rapidly learn to avoid them. Having tasted one and perhaps vomited (as our jay does on page 31), he may sample another before he learns his lesson. But he will avoid them thereafter, especially if the unpleasant experience is fresh in his memory. Thus the more the larvae cluster, the better the chances of encouraging the jay to learn avoidance. And their conspicuous coloration will provide eye-catching, easily identifiable cues for a learned response. The bird will even avoid palatable species that have evolved similar cues.

The European carrion crow preys upon the eggs of black-headed gulls, whose nests are clumped. From what we have seen so far of the strategies of prey, we might assume that gulls' eggs do not taste very good. But I have found them delicious, and I suppose that crows have too. This seeming violation of the "bad-tasting prey clump together" principle can be explained by another strategy available to prey: Gull parents, by living close to each other, cooperate in fighting off predators, as Niko Tinbergen's studies show.

The gulls prefer colonial nesting, and experiments demonstrate that more young survive in nests toward the center of the colony than on the edge of it. The nests are not arranged at random but neatly spaced out. The gull vigorously defends a small territory around its nest. The spacing discourages the black-headed gulls' tendency—unusual among animals—to prey on each other's eggs and young. And, when a fox or other predator prowls the gullery, the spacing of eggs and chicks, as well as the attacks of irate parents, may deter him from specializing in gulls as a source of food.

Another gull species, the kittiwake, must adjust its behavior to extremely crowded living conditions. Unlike most gull species, kittiwakes nest on the narrow ledges of high cliffs. Their choice of such safe nest sites explains their (Continued on page 40)

TRICK OR TREAT?

A bird's-eye view of this caterpillar is alarming: Large round spots on its back look like protruding eyes, even to fake highlights in the black "pupils." Eyespots, an adaptation used by some insects to frighten birds, serve this spicebush swallowtail larva. Earlier in the larval stage the swallowtail also relies upon green camouflage, another defense in the bag of insect tricks. Spacing and cryptic coloration counteract a hunting bird's "searching image," a mental picture of a type of prey and a way to catch it. The image is strengthened by frequent captures and reinforced by the reward of tasty food. But as the bird starts confusing green leaves with sparse insects, it gets no reward for its labors. The image of the food that fooled it fades, and the bird begins tuning in to a new prey.

37

Schemes for survival

In a world of eat or be eaten, a theme of survival is fool or be fooled. The rare Malayan mantis (below), a voracious insect-eater, mimics pink orchids and dines on nectar-seeking prey. Its disguise, in turn, thwarts detection by pursuers—birds and lizards that see it as a flower: leg flanges as petals, the green margin of its prothorax as part of the stem, brown markings on its feet and abdomen as wilted areas. At times the mantis may even rock in imitation of a breeze-blown blossom.

Some potential insect prey ward off hunters more directly. The saddleback caterpillar (left) bears an eye-like marking that probably serves as warning coloration, the opposite of cryptic coloration. This tactic attracts a predator's attention, warning that the food it is about to eat may be distasteful or poisonous. The bird that attacks this venom-spiked larva gets conditioned to avoid it—a form of learning like the one involving the jay on page 31.

Plants that rely on insect pollinators usually lure them with color and scent, paying off with nectar. But flowers can fool. A few smell like rotten meat; they hoax carrion-eating beetles, which get no reward. The *Ophrys* orchid (opposite) resembles a female wasp—and even emits a female odor that attracts a male. Trying to mate, he picks up sticky pollen masses that brush off on another orchid.

EDWARD S. ROSS. UPPER: ROSS E. HUTCHINS. OPPOSITE: OTHMAR DANESCH

peculiarities: a tameness and indifference to predators and the relatively uncamouflaged appearance of their chicks. They are also casual about eggshells, predator attractants that other species meticulously carry far from the nest after hatching.

A young kittiwake has no place to flee when aggression breaks out. Instead, it holds its ground and adopts a special posture, hiding its beak in its breast and exposing a black patch on the back of the neck; the sight of this patch inhibits attacks by other kittiwakes. Unlike other gulls, which wrestle opponents to the ground, kittiwakes use a different technique. They try to twist adversaries off the ledge.

Because kittiwake chicks cannot wander from their precarious nest sites, their parents can always find them. Young gulls of other species, which can move around much more, may be killed if they happen to run up to the wrong adult. Several studies of ground-nesting laughing gulls, terns, and gannets have shown that parents and young learn to recognize each other by appearance and by consistent — but slight — individual differences in some of their calls. Though kittiwakes lack this ability and do not need it, the chicks do learn to identify one another. Thus they are able to repel young neighbors so that each family remains true to its own nest site.

In tropical rain forests, mixed flocks of birds — often several species of tanagers and honeycreepers — roam together and show every sign of coordinating their activities. Such an arrangement would seem unnecessary in a place where insects are abundant. But in a flock the members serve as beaters for each other, raising insects much as human beaters raise hares in a hunt. Predatory hawks and eagles prowl the skies above the forest, and the many pairs of eyes in a flock are more alert than a single pair. A mixed flock produces an alliance of species traits that broadens the range of alertness.

Membership in such groups is so advantageous that some habitual member-species have evolved plumage patterns similar to others in the group. And certain species, usually with a "neutral" or nondescript coloration, which other species seem to find inoffensive, apparently attract these birds of different feathers that flock together.

To gain protection against predators, members of a group must possess some means of signaling the presence of danger to one another. They need an alarm call that will alert all members and perhaps send them rushing for cover. But the very act of giving a danger signal places the signaler itself in peril. For if companions can perceive the signal, then so can the predator. This dilemma has led to a fascinating adaptation in the alarm calls of some bird species.

Twenty years after my study of chaffinches in Madingley Wood, I still recall the day I experienced the frustration of a predator seeking a bird. On a limb crouched a male chaffinch, feathers sleeked. Overhead swooped a sparrow hawk. Then I heard a high, thin, drawn-out whistling call, which always seemed to come from somewhere else. After a good deal of fruitless searching for another bird, I gradually realized the male chaffinch I was looking at was a ventriloquist!

Similar ventriloqual calls from a great tit behind me added to my confusion. Then it occurred to me: If I was confused, why not the hawk? For a hawk to make use of such an alarm call in capturing prey, it must first locate the source. So an effective alarm would be one that was difficult to locate. After studying the problem, I concluded that such a sound could be designed; the ideal one would be a thin, high-pitched whistle — just like the birds' alarm calls that I had heard.

We can best appreciate the delicacy of this adaptation when we realize how a bird or mammal locates a sound source by comparing loudness and timing at the two ears. If the sound is louder in the left ear than in the right, and arrives at the left ear first, the hearer looks to the left for the source of the sound. A stereo system works on the same principle by producing slight variations in loudness, quality, and timing from the two speakers.

The easiest sound to locate is a repeated clucking or chipping. The timing of the clucks can be compared accurately at the two ears. Furthermore, the head "blocks" sound in a certain pitch range, so that the cluck will be fainter at the ear farther from the source. Sometimes birds, instead of hiding from a hawk, mob the predator, their cries indicating the enemy's position to their other companions. For this tactic they need an easily locatable call — and mobbing calls of many birds have the necessary clucking or rattling sound.

The chaffinch at right is merrily singing. If he engages in mobbing and we record his call, we can get a picture of the cluckings in a "voiceprint" or sonogram. The two on the left in this display reveal the sound pattern of a mobbing call: short, quickly rising cries.

If a sound begins and ends gradually, with no sudden breaks or changes in loudness, the difference in the time of arrival to each of the listener's ears will be difficult to determine; if the pitch is pure, in a range where the head does not block it, such a call will be truly ventriloqual. In a sonogram, as we can see here, the call appears as a long, high-pitched sound. An expert in Oriental music pointed out to me that what the Chinese term a "floating note," played on a stringed instrument, has much the same structure as the chaffinch hawk-call.

ERIC HOSKING

Many small birds and even some mammals such as squirrels and marmots use a call of this type, which transmits alarm not only to their own kind but also to other species. To communicate danger efficiently, alarm signals of various bird species should be as similar as possible. But in a situation such as breeding the difference between species must be signaled. That is why mating songs are so different from one another, often giving the bird-watcher — like the birds he is watching — the most reliable clues for identification.

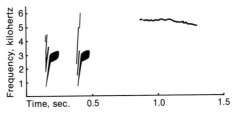

Perhaps the most important opportunity provided by membership in a very close-knit group is the chance to learn from one another. We tend to think of such culture as a uniquely human attribute. But many animal societies also possess it.

We have no more than an inkling of the myriad ways in which young animals learn from their parents, siblings, and other companions. So far, only elementary examples have been analyzed. Domestic chicks will rush to their mother when she pecks at grain.   41

A tradition flows away as a female macaque washes the sand off a sweet potato and eats it. Fed by Japanese scientists at an island research center, monkeys got the habit of thrusting hands to men for handouts. Dunking inspired new behavior. Some monkeys ran 30 yards to the sea, hands full of food. From these burdened runners comes a hint of how man's ancestors began walking on two legs.

Do they learn anything from her rejection of some objects in favor of others? A simple experiment gives an affirmative answer. Chicks are shown a cardboard silhouette of a mechanical hen manipulated so that it pecks at green- or orange-colored grains. In a demonstration not only of learning but also of "contagious behavior," the chicks will immediately run and preferentially peck at that color grain chosen by the mother-model.

Birds placed in separate cages will watch one another attentively. When one bird sees what another eats, the observing bird will move to the feeding area in its own cage and make a similar selection, establishing a food "tradition." Presumably this is how the habit of raiding milk bottles spread among birds in England.

Japanese macaques eat many different plants, but each troop has its own specialties. A yearling learns what to eat by gathering scraps dropped from its mother's mouth. If a youngster picks up something on the forbidden-food list, its mother snatches it away. By the age of 2 most young restrict their food to the troop's traditional diet.

Since 1952 an island troop of macaques has been provisioned with sweet potatoes by Japanese scientists at a feeding station on the beach. One day Imo, a 1½-year-old female, was seen washing the sand off a sweet potato. This new tradition was first imitated by one of Imo's playmates, then by Imo's mother. The potato-washing behavior continued to spread, generally from youngsters to mothers and from younger brothers and sisters to older ones. At this time, members of the troop did not bathe in the sea.

Four years later the only adult animals that had learned to wash potatoes were mothers of potato-washing youngsters. Then, when playmates became mothers, they waded into the sea to wash potatoes with their young clinging tightly to them. The offspring imitated their mothers as if monkeys had always washed sweet potatoes.

The macaques also eat wheat that is scattered on the beach and becomes mixed with sand. The members of Imo's troop laboriously scoured the wet sand for single grains of wheat. Then Imo, at age 4, was seen to pick up a handful of sand and wheat, toss it into the sea, and scoop up the floating wheat after the sand had sunk. In the next six years most monkeys started wheat-sifting when they were 2, 3, or 4 years old. Adult males of the troop, including those of the highest rank in the dominance hierarchy, usually do not enter the sea—even after their favorite food, peanuts. Here we see the other side of the story: Animals bound by tradition resist the development of new behavioral patterns.

We tend to think of the ability to modify behavior as the acme of adaptability. We assume that an animal lacking this flexibility does not have the complex nervous system that learning requires. But sole dependence upon learning can be hazardous; an animal that has to learn every source of danger is likely to be dead before the process is complete. Learning left to the mercy of accidental events more likely generates superstitious rather than adaptive behavior. This is what Konrad Lorenz had in mind in proposing the "innate schoolmarm" as a necessary prerequisite for adaptive learning.

We can see the schoolmarm at work in the development of birdsong dialects. Like many songbirds, a male white-crowned sparrow must learn the song of his species. If a male is raised in a laboratory out of hearing of adults of his own kind, his song will develop abnormally. But if we play recordings of normal songs—at some time between 10 and 50 days of age—he not only will learn to sing normally but also will imitate the particular local dialect he hears. (In this species' song, dialects are analogous to those in human speech.) If during this period we play him recordings of another bird—even such a close relative as the song sparrow—he will not learn its song.

Thus, although the sparrow cannot produce the normal song without training, he does have some internal guidelines that focus his attention on a certain kind of example. He carries insurance that the learning process will not go completely awry and lead him to learn another species' song. The existence of such learning guidelines may prove to be important in many other species, including our own.

Human language is a biological phenomenon with an evolutionary history. Linguists who analyze the deep structure of languages and psychologists who study the earliest

Stealing sips from milk bottles they broke into, two blue tits carry on a larcenous tradition first noted in England around 1921 and later on the Continent. About a dozen species learned the crime, a few as witnesses. Some birds await milkmen at houses where people are slow learners and don't box bottles.

RONALD THOMPSON, BRUCE COLEMAN INC. OPPOSITE: C. RAY CARPENTER

With a tender touch of hands two chimpanzees cement social bonds and give assurance of nonaggression. Animals often reduce tension by physical contact; some greet one another with an embrace, hold hands, or pat a subordinate, much as a mother calms her baby. Social grooming especially tends to pacify a companion, inducing relaxation, stupor, or even sleep. Some animals, such as these black-tailed prairie dogs, use kisses to establish the identity of community members. Sea lions —and chimps—kiss as a greeting. The friendly behavior of a nuzzling pair contrasts with that of jealous male rivals. Old bulls charge any animal that challenges their authority as harem-masters. Mammals and birds generally attack companions that crowd them. But intruders of the opposite sex are allowed to come closer before hostilities begin.

ROGER TORY PETERSON. ABOVE: J. PERLEY FITZGERALD. TOP: HUGO VAN LAWICK

speech patterns agree that children must have internal guidelines that favor certain kinds of grammar. How otherwise to explain that, in spite of all the surface differences between languages, all men speak according to the same basic rules?

Like some birds, children do not approach vocal learning with the mythical blank slate on which experience can write absolutely anything. They are predisposed to learn more readily in some directions than others and at certain critical times. Isolated birds learning songs by the playback of sounds seem to need no reward of food, water, or social stimulation—"reinforcements" that some psychologists say children need. Likewise, children matching the sounds of their own vocalizations with the sounds they hear, seem to need no direct rewards from their parents. We may be groping toward a basic set of signal-learning rules for any species that needs guidance in learning. And these rules may be ones to which even our own species is subject.

Through learning, an animal acquires specific habits that teach it what to eat and what to spurn, what to sing and whom to mate with. At a deeper, emotional level, an animal also acquires habits that lead toward relaxed, pleasurable companionship with some individuals—and tension and antipathy toward others.

In a rapid learning process called "imprinting," ducks, geese, and some other birds will follow soon after birth whatever moving object satisfies certain requirements. As they follow, they learn the object's characteristics and so develop a bond that may endure into adulthood. An object that makes sounds is more readily adopted as a parent surrogate than a silent one; a young wood duck, say, will usually choose to follow an object making the sounds of a wood duck mother. The appearance of the mother, and of a mate later on, must be learned. But by being predisposed to the species' sound, the duckling is less likely to accept as a mother—and later, as a mate—a duck of the wrong species.

We know from the work of the distinguished primatologist Harry Harlow that if an infant rhesus monkey is separated from its mother and siblings, even for only a few weeks, its temperament is radically changed. This behavior bears many resemblances to a depressed condition known in children raised in institutions. Studies with experimental animals have thus given us good reason to think that relatively brief separation from his mother and playmates can emotionally injure a child.

Gene Sackett, a colleague of Harlow, took rhesus monkeys from their mothers at birth. He showed them slides of monkeys, some in relaxed poses and others displaying gestures and expressions of aggression. Though they had never had contact with other monkeys, the social isolates strongly responded to slides showing the typical "open-mouthed threat" of a higher-ranking animal about to attack. Of all communicative gestures, better to be born knowing this one than be injured while learning it.

In a lasting social relationship, animals must learn the signs that serve to maintain and cement the bond of friendship. Reduced to its simple evolutionary requirements, the social bond needs two links: The companions should engage in some mutually pleasurable (and thus most likely repetitive) activity, and they need ways to reduce the chances of fighting and to ameliorate or divert aggression.

Sometimes the same activity, such as food-sharing, may forge both links. Since the actions of food-sharing are physically incompatible with threat and combat, they may also serve to discourage fighting. Postures of submission, seemingly designed to inhibit an impending attack, often include miming of food-sharing or sexual acts. An animal peacefully seeking to inhibit attack avoids any act that may be construed as aggressive. The direct frontal gaze at the opponent in an aggressive display contrasts strikingly with the shifty averted gaze of a submissive animal—which also usually avoids vigorous movements and, as a gesture of nonaggression, conceals its weapons.

We know, through modern techniques of group therapy, the value of physical contact as a method of reducing tensions. Many nonhuman primates use social grooming in much the same way. Once we appreciate the potential value of bodily contact, we begin to understand the significance of such primate behavior as the hand-touching and kissing of chimpanzees. These gestures often have human overtones, and even the most cautious behaviorist finds it difficult to refrain from empathizing with them.

Strange surroundings arouse and disturb animals. Given the chance, they will seek out familiar ground. The strongest stimulus for aggression in animals as diverse as fish and apes is the introduction of a stranger of the same species.

Could it be that familiarity is a kind of social cement? Perhaps social grooming furthers such familiarity through all the senses. If we assume that frequent intimate contact— exchanging food, sexual activity, even simply resting closely together—is a need, an animal would more easily find such contact in its own community than in a strange group. We can imagine a maximum group size within which individuals could establish and sustain sensory familiarity; the ideal would vary from few to many, according to the species.

To the concept of an ideal group size we should add that lack of adequate space is a basic stimulus for hostility. Aggression may also relate to a particular place, as when a territory is defended. These predispositions for aggression can be enhanced or subdued by learning, within the guidelines set by the species.

Our own species most likely has such predispositions. This is not to say that aggression is as inevitable and inexorable as the growth of an arm or a foot. Everything we know about animal aggression suggests that it can be muted, modified, or even inhibited by the right kind of social interchanges. The challenge to humanity is to discover just what our predispositions are. Armed with that knowledge, we can hopefully design social contacts to minimize the triggering of aggression.

At the level of small gestures between individuals—tones of speech, modes of address, the social courtesies which we so impatiently overlook in the fast pace of modern life—the threshold for aggression may be raised or lowered. If we deny ourselves those so-called small gestures, we are much more likely to overreact under some sudden social pressure, perhaps breaking out as a group into one of those shameful acts of murder and destruction that are so much a part of our times. Animals neglectful of the rules face the prospect of living as social outcasts. Perhaps in those rules of animal behavior there is a message for us, if only we will learn to listen.

Deprived of a mother of flesh and blood, a rhesus monkey clings to its choice—a terry-cloth proxy that offers only a texture to touch. The infant rejects a mother that offers milk but whose wire feels wrong. Touch lessens terror; soon the monkey will explore its world, gaining confidence to bridge the chasm between security and food. Holding to one mother, it will stretch to nurse from the other.

MARTIN ROGERS

Aubrey Manning

THE WORK OF BEING A BEE

Sallying from the hive at the first blush of spring, the honeybee searches out the nodding blossoms that break the siege of winter. Pausing here and there—one among thousands of sisters out foraging—she collects her tiny contribution to the food stores of the hive. She may visit hundreds of blooms on an hour-long sortie, sipping sugary nectar with her tube-like proboscis or brushing protein-rich pollen into bristly "baskets" on her hind legs. Returning to the hive, she is met on the combs by other workers. They relieve her of her burden and she flies away. Again and again she repeats her mission until lengthening shadows and the sun's waning rays bring an end to her day in the field.

The busy forager is but 3 weeks old, and she has less than another month to live. Then, exhausted by the demands of her task, her wings in tatters, she will leave the hive to return no more. The food she has gathered will feed other generations in the colony.

Bees can be found wherever flowers bloom in profusion, for the two kinds of life—one animal, one plant—form one of nature's striking examples of mutual adaptation. Evolving together through the millenniums, many flowers have come to depend on bees for pollination, and bees to depend entirely upon flowers for food.

Honeybees live in highly organized societies that enable them to exploit flower crops efficiently and to control precisely the living conditions within the hive. Though *Apis mellifera* is a common sight on a midsummer day, it is only one species of a very large family of bees, the majority of which live solitarily.

We do not know exactly how social life began among bees, although it most probably arose with a solitary female constructing several adjacent brood cells and living long enough to be still at work when her offspring emerged. The daughters may have begun to cooperate with their mother to build more cells. We could then call the original female the queen and her daughters the workers.

Gathering pollen from a willow bloom, a honeybee savors the fullness of spring. She moistens the golden spores with her tongue, then combs them into pellets which she stows on her legs to carry back to the hive.

Honeybees are thought to have originated in India, spreading to Europe and Africa. For thousands of years man has kept their hives —and pondered their society. "The bee is more honored than other animals," preached St. John Chrysostom in the fourth century, "not because she labors, but because she labors for others."

Unknown to the New World before colonial times, honeybees may have been brought to Brazil around 1530 and to North America about a century later.

Among the world's myriad insect species, some 20,000 are bees. Of these, fewer than 5 percent lead social lives.

HERMAN EISENBEISS, PHOTO RESEARCHERS

From this evolutionary stage it is not a big step to the more advanced social system of the familiar bumblebee. Although a bumblebee queen begins life in the spring as a solitary worker, she soon raises a colony of sterile daughters that cooperate with her to enlarge the nest. Eventually the queen stays permanently in the nest, laying eggs in cells prepared for them, until at the end of summer the colony may number several hundred. By then, stored food is plentiful and some large female bumblebees are produced, as well as a few males, or drones. After mating, the males die. By late autumn the sterile workers are dead and the fertilized queens have burrowed into the ground to hibernate through the winter.

The high point of social evolution among bees is achieved by honeybees. They no longer have a solitary stage in their life cycle; a honeybee queen is so specialized she cannot establish a colony without the help of several thousand daughters. Colonies are perennial—some persist 20 years or more in the wild—because the worker teams of summer can amass food surpluses large enough to keep their hivemates fed in reduced colonies through the winter.

Wild honeybees usually build their combs in hollow trees or in the dark lofts of barns. Each comb consists of a wax base that hangs vertically with a layer of cells projecting horizontally on each side;

two or three combs may be built parallel to each other. Cells in which the larvae are reared—the "brood chamber" area of the comb—tend to be near the center. Pollen and the honey made from nectar are stored separately in cells near the top. Nurtured by pollen protein for development and honey for energy, the young grow quickly. By the middle of summer, at the height of the nectar flow, the number of workers in a single hive can swell to 80,000.

Two factors make possible the elaborate social organization of the honeybee and other social insects such as the ant (page 127). One is division of labor within the colony; the other, communication between its members. Physiology, of course, plays a decisive part in determining work roles. Queens and workers are females, but since a worker's ovaries normally remain undeveloped, she is sterile. Her stinger is actually a modified ovipositor. A queen is not much larger than a worker in the head and thorax, but, with a much larger abdomen, she is specialized for egg production. Conversely, the queen's mandibles are the wrong shape to build wax cells and she lacks the stiff hairs that form the worker's pollen baskets.

By itself this division into castes is not sufficient to account for the many different jobs that bees perform. If you watch the workers,

Solo worker: A primitive solitary bee of the genus Megachile *snips petals from a* Clarkia *blossom to build a nursery for offspring she will never see. Working swiftly, the half-inch insect rolls the cut petal into a tube and carries it to her burrow; the round-trip flight may take only ten seconds. Completing the chamber that will serve as a cradle through the winter, she packs one end of it with beebread—a nutritious mixture of pollen and nectar—and on it deposits a single egg. She plugs the tube with more cuttings and, task completed, flies away to build anew. In this manner the bee constructs half a dozen or so nest cells during a brief life geared precisely to the blooms of a single plant species.*

The bumblebee (opposite), more socially advanced, may live in a colony consisting of several hundred members. Here a worker tends a larva in its waxen cell, feeding it pollen and honey through a hole in the top. When the larva matures, it spins a cocoon and the purplish wax cover is stripped away to be used again. Unlike the Megachile, *bumblebees work cooperatively and gather supplies of food from a wide variety of flowers.*

Solicitous workers encircle
a honeybee queen as she pursues
her destiny: laying eggs that
will assure her colony's survival.
Clipped wing and a daub of nail
polish on her thorax help
the beekeeper find her quickly
amid the hive's milling masses.
The queen mother labors day
and night throughout spring and
summer, pausing only to be fed
or groomed by her attendants.
At the crest of her reproductive
surge she may lay 1,500 eggs
a day—more than her body weight!

Wondrously efficient at her
task, she traces a spiral path
from the center of a comb,
methodically inspecting each
empty cell before she deposits
an egg in it. Larvae hatch in
3 days and are dutifully tended
by nurse bees, specialized
workers too young for the field.
Each larva, feeding voraciously
over the next 6 days, gains up
to 1,500 times its birth weight.
Then it is sealed in its cell,
to emerge fully grown 12 days
later—21 days from egg to adult.

Honeybee queens, usually born
in late spring, reach adulthood
after only 16 days. The first
to break from her cell normally
destroys any sister queens.
The queen mates with a few drones
during a series of nuptial flights,
storing enough sperm to fertilize
the hundreds of thousands of
worker eggs she bears throughout
the 3 or 4 years of her life.
EDWARD S. ROSS

it soon becomes obvious that their behavior is consistent and that they do not shift aimlessly from task to task. One bee may spend an hour readying cells for fresh eggs; another moves across the combs to attend the queen; a third forages in the field. How does each worker know precisely what to do and when to do it?

Half a century ago the German entomologist Gustav Rösch, in a series of classic experiments, determined that what a worker does depends largely on her age—that the sequence of tasks she performs throughout her brief life is dictated by fundamental changes within her body. For example, a pair of "nurse" glands in her head secrete food essential to the development of young larvae. The substance, known as bee milk or royal jelly, is fed to all larvae for three days after they hatch and thereafter only to queen larvae. Worker larvae are weaned to diets of pollen and honey. Nurse glands develop when a worker is 5 or 6 days old, but until then she cannot feed the youngest larvae. Similarly, the decline of her feeding duties—and the onset of comb-making and repair—coincide with the growth of abdominal wax-producing glands, which mature when she is about 12 days old. A week or so later the wax glands wither and the worker begins the life of a forager.

But age alone does not always determine a worker's duties. If it did, a colony would be unable to contend with emergencies or with vagaries of weather and harvest seasons. Honeybees are remarkably flexible; they can change tasks whenever necessary.

Rösch once deprived a colony of its older workers, the foragers. Some of the younger "house bees" stopped their hive duties and took to the fields to forage. Conversely, when Rösch later removed the nurse-workers, many of the foragers began to tend the brood again, even regenerating their nurse glands for the purpose!

Exactly how does a bee detect a shortage of food, of brood nursers, or of new cells? By observation and communication. A worker spends much of her time wandering about the combs, exploring the empty cells, the food stores, and the brood area. By doing so she probably obtains firsthand information about hive needs.

Communication takes place when a patrolling worker meets another worker. Colony members share their food, begging it or offering it whenever they meet. Most food probably passes from incoming foragers to the house bees, but even among house bees there is much exchange of regurgitated nectar. In one experiment by British scientists, 6 foragers from a colony of 24,600 bees took back to their hive a sugar syrup containing radioactive phosphorus. By the next day more than half the house bees and three-fourths of all the foragers had traces of radioactivity in their crops.

Breaking from his larval prison, an adult drone peers big-eyed at the honeybee world about him. Males, born from unfertilized eggs, lack structures needed to carry on the work of the hive. Maligned as sluggards, they are utterly dependent on the bounty of their industrious sisters. Nature denies males even a weapon of defense, the stinger. Their sole function—to find and fertilize a young queen— is enhanced by compound eyes that see in nearly all directions. Drones die after the mating act; those that fail in their mission are cast out of the hive to perish when autumn sets in.

Pendant combs, heavy with food and wormlike larvae, form the heart of honeybee endeavor (opposite). Workers at top dip into stores of pollen and honey to feed young in unsealed cells at the edges of the central brood area. Half a dozen others clean and repair empty cells, including slightly larger drone chambers below the main brood.

STEPHEN DALTON. OPPOSITE, UPPER: © WALT DISNEY PRODUCTIONS. LOWER: EDWARD S. ROSS

Foragers pausing by a shaded pool take on water needed in the hive. In spring, when brood cells bulge with ravenous larvae, nurses in a large colony may use a pint a day to dilute thick, stored honey before feeding it to their charges. On hot days, workers often collect large amounts to help cool the hive.

Her head serving as a ramrod, another worker (opposite, upper) tamps pollen into a storage cell. To preserve the perishable load, she tops the cell with a layer of honey, then seals it with an impervious wax cap.

Whirring wings fan a trail of scent for returning hivemates (lower). Flexing the tip of her abdomen, a worker reveals the lip-like Nassanoff gland she uses as a scent beacon to help guide nearsighted companions to the hive entrance or to flowers she discovers abroad.

Scientists have found that a bee's duties are determined largely by her age. The table below summarizes her busy life:

TASKS	ADULT AGE (IN DAYS)
Cleaning cells, incubating brood	0-4
Feeding older larvae	4-6
Feeding both younger and older larvae	6-12
Building and repairing combs; storing food	12-21
Foraging	21-42

Food-sharing not only imbues the bees in a hive with a distinctive odor that makes each readily identifiable as a member of the colony but also provides vital information about the queen. As she moves about the comb, workers cluster around her in a small group, plying her with food, stroking her with their antennae, and licking her with their tongues. By licking the queen the attendants ingest palatable body secretions called queen substance. Although only a few dozen attendants come into direct contact with their queen, the practice of sharing food with other workers quickly broadcasts minute amounts of the substance throughout the colony. In effect the message is conveyed that the queen is alive and well.

If the queen dies, the amount of queen substance shared by any given bee falls below a critical level. Within an hour or two of her death the colony mobilizes to build special queen cells and to feed several larvae the continuous diet of royal jelly that will assure a fertile successor. Similarly, when a hive becomes overpopulated, the secretions of a single queen become too dilute. This may in turn trigger a response to prepare for a new queen so that part of the colony can swarm to a new site.

Food-sharing and direct contact also help foragers keep a close check on the hive's food supply. If food is scarce, house bees beg eagerly from a returning forager; but if food is plentiful or if her nectar is watery, she may have to proffer it repeatedly before she can find a house bee willing to take it from her. Thus the forager's

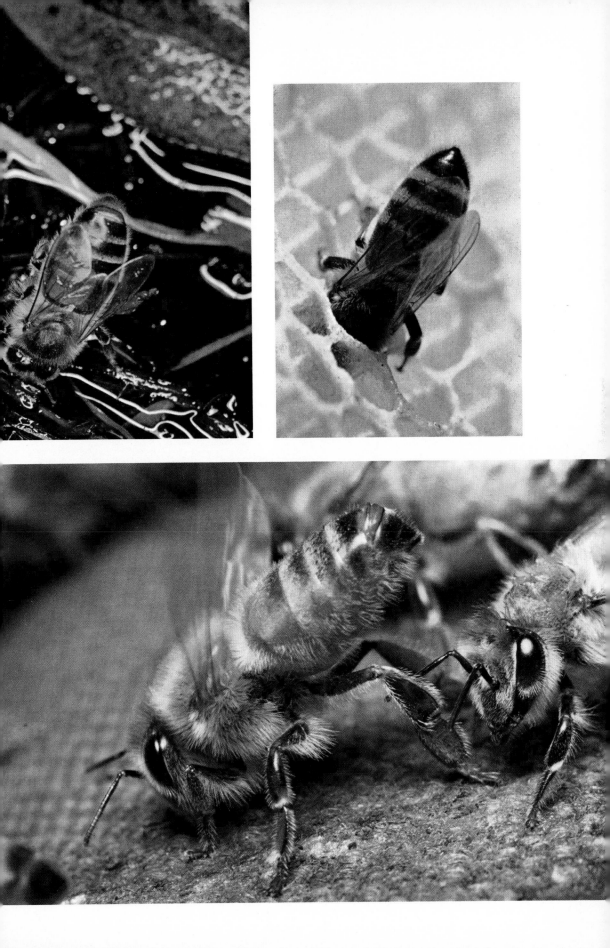

reception in the hive indicates to her whether to collect more nectar, switch to a better crop, or stay in the hive and rest. If the forager is on a pollen-gathering mission, she can tell whether more is needed by examining the storage cells.

One of the most remarkable examples of how foragers adjust behavior to fit conditions can be seen when the hive begins to overheat. Whatever the temperature outside, honeybees maintain the hive — especially the brood chamber — between 92° and 95° Fahrenheit. In the chill of early morning the workers cluster over the comb, incubating the brood with body heat. As the day warms, the tightly knit cluster gradually disperses. If the temperature continues to climb, some of the workers begin to fan their wings, directing a cooling flow of air through the hive entrance and over the combs. On an extremely hot day, the bees must resort to an even more drastic cooling measure: They place drops of dilute honey at the openings of empty cells, evaporating the liquid in the airstream created by their wings. If there is not enough fresh, unconcentrated honey on hand, the foragers begin to scout for water.

Here again, the signal to switch from one activity to another is communicated by the behavior of the hive bees. Loads of rich, concentrated nectar are refused repeatedly; only very dilute nectar is accepted. This unusual situation, perhaps coupled with their own perception of the hive's warmth, alerts the foragers to begin collecting water — a task they pursue until the crisis has passed.

By far the most elaborate communication system — one almost without parallel in the animal world — exists among a hive's foraging

Tongue uncoiled, a honeybee probes for nectar amid clover petals. When the nectaries of the plant are located, other mouthparts swing down; they form the proboscis through which she sips. The tongue, about 3/16ths of an inch long, tapers to a spoon-shaped tip.

For at least 60 million years bees and flowers have lived in symbiotic partnership. The first flowers, simple open structures with exposed stamens, probably were pollinated only by wind and beetles. Some gradually became more specialized, luring insects with colors and fragrances — and conveniences such as the landing-platform petals of the sticky sage (left). As the bee pushes in, anthers droop, dusting pollen on her back.

Bees and many other insects that depend on flowers for food developed specialized mouths and internal storage sacs, or crops, that hold the nectar without digesting it. A bee's penchant for visiting only one type of flower on any given mission helps ensure that pollen she inadvertently collects on her body will propagate other plants of the same species.

HERMAN EISENBEISS, PHOTO RESEARCHERS
OPPOSITE: TREAT DAVIDSON

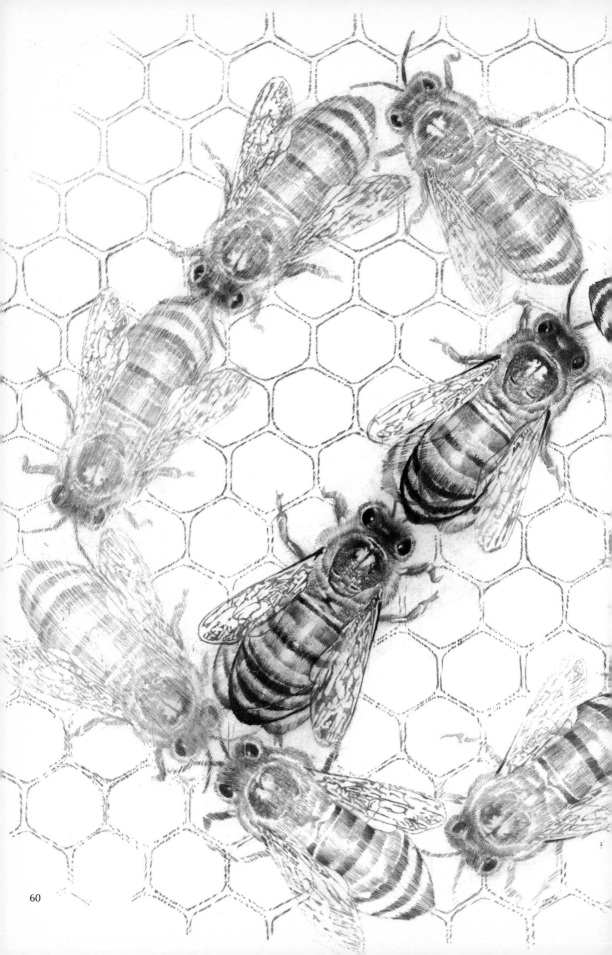

Sun dance of the honeybee: messages in motion

Within the gloom of the hive, a honeybee traces a series of figure 8's across the face of a comb. On the straight part of her run—between the loops of the 8—the dancer buzzes noisily and may shake her entire body with a violence that suggests frenzy. The curious performance, known as the waggle dance, is actually a remarkable method of communication through which the honeybee conveys information—the range and bearing to a rich new food source. In this multiple-image drawing, the dancer circles again and again, first to the left, then to the right. Other workers (not shown) cluster around her, following her every move intently, their antennae pressed against her vibrating abdomen.

"The idea of distance," writes Dr. Manning, "is imparted by the speed of the dance; the more 8's the dancer completes per unit of time, the closer the food source. Thus, if the target is 200 yards away, the bee runs eight or nine loops in 15 seconds, but only four if the source is 1,500 yards distant.

"The dancer may also convey distance by the number of waggles she makes on the straight part of her run, by the length of time she spends on each run, or by the duration of the buzzing that accompanies it. We are not sure which of these codes—sound or tempo—the other bees respond to. Quite possibly they respond to them all."

Different races of bees have different "dialects." Austrian bees, for example, cannot correctly "read" a distance message of slower-moving Italian bees, though both use similar figure-8 dances.

To convey direction, the bee angles the straight part of her run to coincide with the angle between the sun and the food source. She uses gravity to represent the sun's position, in effect "translating" the solar bearing from a horizontal plane to a vertical one on the face of the comb. On a breezy day, she even corrects for windage. In the drawing, as in the diagram below, she indicates a course 40 degrees to the right of the sun by angling her run 40 degrees to the right of vertical. To indicate a course, say, 70 degrees to the sun's left, she would shift her run correspondingly to the left.

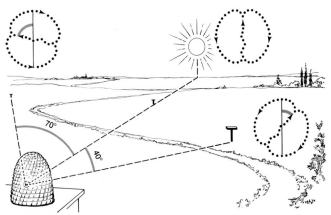

DRAWING BY GEORGE FOUNDS. ABOVE: LISA BIGANZOLI, GEOGRAPHIC ART DIVISION

bees. Anyone can observe its effects simply by placing a dish of concentrated sugar water in the garden. Several hours may pass before a bee finds the prize, but eventually one will. She fills her crop with the sweetened solution and promptly returns to the hive. Within minutes there will be a rapid buildup of bees around the dish — whether or not the original finder returns. Obviously the first honeybee to find the dish has imparted her discovery to the colony's other foragers. But how did she do it? And how did she let the others know precisely where to find the rich new food source?

By dancing! In what ranks as one of the great biological triumphs of the century, Austrian zoologist Karl von Frisch in the early 1920's discovered the significance of the honeybees' dance. He found that if a new source of food is located within about 100 yards of the hive, the finder dances in tight little circles on the face of the comb, often reversing direction before completing a full revolution. She may

dance half a minute or more, then move to another part of the comb and repeat it. This round dance, followed by other foragers in the hive, stimulates them to search at random, but close to the hive. Flower scent clinging to the dancer and the sight of other bees near the food source help the foragers pinpoint the target.

A different dance, called the waggle dance, is used to indicate a food source more than 100 yards away. Now the dancer traces on the comb a path resembling a compressed figure 8. Again and again she repeats the stereotyped pattern, and on the straight part of her run she vigorously shakes her body, particularly her abdomen. Throughout the dance other workers follow her every move, keeping their heads close to her body, touching her with quivering antennae. Von Frisch discovered that foragers leaving the hive after following a waggle dance do not search at random; they tend to fly directly to the food source. Through elaborate experiments, he

Abdomen swollen with riches of the field, a homecoming worker proffers nectar to a hivemate that extends her tongue to solicit a sip. A forager may visit 50 to 800 blooms on an hour's sortie, returning with a cargo that weighs up to 85 percent as much as she does. Her life's work yields but half a teaspoon of honey.

To convert nectar to honey, the hive bee first finds a quiet place on the combs. She then regurgitates a droplet of nectar, holding it briefly between jaw and tongue to let excess water evaporate. She swallows the drop and brings up another, a process she repeats 80 or 90 times in about 20 minutes, or until her entire load has been exposed to the drying air of the hive.

The bee also adds invertase, a digestive enzyme that breaks complex sugars into simple ones readily digested by both bee and man. Stored in open cells to thicken by further processing and evaporation, the syrup ripens into amber-colored honey.

A colony of honeybees living in the wild may make between 30 and 60 pounds in a season; domesticated ones, with man's help, produce up to 200 pounds.
EDWARD S. ROSS

found that the speed of the dance (the number of loops completed per unit of time) correlates closely with distance to the food source. Thus, the closer the food, the faster the bee traces her 8's. Recent studies indicate that the dancer may also signal distance by the number of waggles on the straightaway or by the duration of the buzzing sound that accompanies them.

The dancer also uses the straight part of her run to indicate direction. She can do this in total darkness on the vertical surface of a comb because of her ability to represent the sun's position in relation to the pull of gravity. Thus, if she runs straight up the comb—"away from gravity"—she tells the others to fly directly toward the sun. If she runs straight down the comb—"toward gravity"—she tells them to fly directly away from the sun. To indicate other directions, the dancer rotates the axis of her run to correspond with the angle between sun and food source (page 61). She finds the sun even on an overcast day because she sees its ultraviolet rays.

While many insects, birds, fishes, and mammals can navigate by the sun, the ability of the honeybee to take solar bearings and to pass along the information is a feat only man seems able to match.

Foragers that collect from blooms with poor yields tend to lose interest after a while and return to the hive to rest. Those that find rich crops tend to dance. The better the crop, the more persistently they dance and the more followers they recruit. Thus the attention of the colony's foraging team is always efficiently directed toward harvesting the best food sources available.

Simple and effective. The two words sum up the nature of a honeybee's social life. Such is the marvel of her adaptability that we may feel she reacts intelligently—much the same way we humans do. But this is not the way the bee works. She is endowed with only a limited set of inherited responses to relatively uncomplicated stimuli which she receives from such sources as the comb, the larvae, the queen, other workers, flowers, and the sun. She is constantly recording changes in her environment and is able, within limits, to alter her behavior to meet each change. From such stereotyped responses, the behavior of many thousands of bees is integrated to form a cohesive colony. The queen relays her genetic characteristics to her daughters. They, in turn, help her rear more daughters and the few fertile queens that will keep the colony viable long after the old queen has departed.

Though honeybees rely on what many call blind instinct, we need not lose our sense of wonder at the remarkable way they manage their lives. Rather, we should marvel the more that such seemingly complex ends can be achieved by such uncomplicated means.

Tiny voyagers, setting forth to establish a new colony, mass on a branch close to their ancestral home. They may cling together for hours or even days while their scouts search for a permanent site. Successful scouts perform waggle dances on the clustered bodies of their companions. The better the find, the more excitedly they dance and the more followers they recruit. After many inspection flights, the best spot is picked and the swarm decamps.

Scientists theorize that hive prosperity triggers the honeybee impulse to move. A few weeks before swarming, usually in spring or early summer, the bees grow restless. Workers in the crowded hive begin to shape new queen cells near the edges of the brood comb and the queen herself may slow or halt egg production. Foragers faced with glutted honeycombs become lackadaisical about their work and take to idling at home. Some become scouts and fly out, poking into nooks and hollows to find attractive nesting sites. Then, on a balmy day before the first new queen emerges from her cell as an adult, the dowager and perhaps half the workers come tumbling out of the hive— an avalanche of whirring wings and tawny bodies confronting the uncertainties of a new world.
STEPHEN DALTON

George B. Schaller

THE SOCIABLE KINGDOM

The male lion stood silhouetted in the day's last light, imperious and unconcealed. Before him five lionesses crouched low in the grass and began to stalk their prey, a herd of wildebeests. The pride had spotted the herd some 45 minutes before, when the wildebeests were more than a mile away, black dots moving across the Serengeti Plain. The lions had followed slowly, casually. They closed in with the darkness, the lionesses fanning out on a broad front while the male remained behind.

Now a lioness on the left flank rushed. Dark shapes wheeled and scattered, most of the herd bolting to the right. But one wildebeest was too slow. A lioness pulled it down and two cats raced to help her. While the other two lionesses and the male vainly pursued the stampeded herd, I hurried toward the site of the kill. One lioness

Lionesses and cubs pause for a sociable drink in Tanzania's Serengeti National Park, heart of African lion country. Their copycat lineup displays a trait of social animals: what one does, others do. Members of a pride, they are all related—mothers, daughters, granddaughters in union for life. Few female cubs leave their family pride; young males depart for a nomadic life.

In his study of the king of beasts, the author learned that the monarch does little hunting, often dines on kills of other predators—and may be unregally driven from a meal by hyena commoners.
GEORGE B. SCHALLER

67

held the downed wildebeest by its throat, another by its muzzle, suffocating it. Then, flank to flank over the carcass, snarling and growling, the pride fought for the spoils.

I had not merely watched all this. Muscles flexed, holding my breath, I had participated. The tension of the hunt—to see lions snake tautly toward their prey, hesitate fleetingly, then explode into dispassionate violence—was a drama of unrestrained vitality. I had felt my body unconsciously reverting to a predatory past, when man too was a hunter.

Perhaps the hunt had told me something about man. But I had come to Serengeti National Park in Tanzania for a three-year study of the lion and other predators. The project was financed by the National Science Foundation and the New York Zoological Society, with photographic help from the National Geographic Society.

I wondered why lions were so intensely social while other cats were essentially solitary. I was intrigued by the cooperative hunting and killing techniques of lionesses, by the aloof role of the male, and by the seemingly excessive aggression at a kill. Traits such as these must have survival value. It was my task to search for the evolutionary forces that had molded lion society.

To try to answer my questions, I observed certain prides in all seasons, day and night. I recorded births and deaths, courtships and fights. I studied other predators to find out where they fit in the Serengeti environment. And I became intrigued by the question of how my work applied to man—the predatory primate.

I was puzzled by the lions' social system until I got to know by sight every lion in the area. I soon noticed that some never associated. Others greeted each other in typical cat fashion by rubbing cheeks; these lions belonged to the same pride, consisting of one to four males and several females and their cubs.

Though all of its members seldom roam together, the pride itself endures for years, and most pride lionesses remain in it for life, generation succeeding generation. Males, however, do not become permanent members. I observed 12 prides for two or more years; only 3 retained the same males. Sooner or later males leave. A few lionesses may also leave and become nomads when they are between 2½ and 3½ years old. Such emigration helps keep the pride's size in proportion to the amount of food available.

A close relationship exists between exploitation of food resources

Lion cubs—born of working mothers and casual fathers—find security in the pride. They play only when they feel safe, testing their prowess in harmless games. This father shows rare affection; a mother, perhaps tired from a hunt, bares her teeth at her hungry nursling. If rebuffed, a cub wanders through the pride until it finds a foster mother willing to suckle it.

UPPER LEFT AND RIGHT: GEORGE B. SCHALLER. CENTER: THOMAS NEBBIA.
LOWER: MARC AND EVELYNE BERNHEIM, RAPHO GUILLUMETTE

Asocial cats, the leopard and the cheetah prefer solitary living. Night-prowling "companion of the moon," the leopard hunts alone, usually hauling the kill —here a Thomson's gazelle— into a tree, wedging the body between branches to keep it from lion and hyena scavengers. The cat feasts undisturbed in its tree three or four days. Serengeti leopards get twice as many male as female gazelles because males are less alert.

The cheetah, built for speed not stamina, streaks in sprints as fast as 60 miles an hour, seldom pursuing prey more than 300 yards. The author recorded 70 percent success in chases by this fleetest of all land mammals.

by a species and its social organization. Most cats probably have a limited social life in part because they hunt small prey. A house cat gains no advantage by sharing a mouse. And, hunting for scattered prey in dense vegetation, a solitary predator possibly has a better chance for success than a pack.

The leopard, a good example of a solitary cat, ranges over 15 to 20 square miles in the Serengeti. Although several females and a male may share a range, adults rarely meet. Cubs begin to roam on their own at just over a year old. But a cub will associate with its mother—and share a kill with her—until it is nearly two. Hunting in riverine forests, along boulder-strewn slopes, and similar habitats offering cover, leopards prey on impalas, duikers, baboons, jackals, and other fairly small animals.

The lions' social system is largely an adaptation for cooperatively hunting large prey in open terrain. The system is flexible, so that pride members may either scatter or hunt together, depending upon the size of available prey. When the gazelle, which provides a meal

for only one lion, is the principal quarry, group size is smaller than when the lions go after larger animals, such as the zebra. But a social existence also has disadvantages. Sometimes competition for food is so fierce that cubs may starve.

To me, the lion's roar is the voice of Africa. On the open plain the night is silent until suddenly a roar fills the void—first a moan or two, then full-throated thunder that finally dies away in a series of hoarse grunts. To a pride member the roar means social contact: "Here I am!" But a strange lion hears a proclamation of ownership: "This land is mine!"

Each pride confines itself to a definite area, varying from 15 to 150 square miles, in which other lions are not readily tolerated. Pride areas overlap extensively, but adjoining groups usually manage to avoid meetings. Males patrol their domain, leaving calling cards of scent mixed with urine on bushes and tufts of grass. Any nomadic stranger or member of another pride that smells the scent knows not only that the area is occupied but perhaps also how recently the owner passed by.

Some pride males willingly abandon their lionesses, but the majority are evicted by nomadic males. A male cub is forced out by

a reigning male, usually by the age of 3½. He becomes a nomad—another of the many extra males that roam around, ready to expel pride males should they show signs of weakness. The nomads often travel in twos or threes; some are brothers that left together.

When two nomads arrive, a single pride male is usually unable to retain jurisdiction over either the pride's terrain or its lionesses. After their overlord has been dislodged, the lionesses readily accept their new masters. Since the lionesses belong to a closed social system, their friendly contacts with strange males advantageously reduce inbreeding in the pride.

Lions generally avoid fights, preferring to escort strangers out of the area. Encounters—mainly male vs. male or female vs. female—usually are nothing more than a flurry of flashing claws and ferocious snarls. But no peacekeeping system works perfectly.

One evening I met the males of the Seronera and Masai prides, as I called them, near each other in an overlapping part of their ranges. At 6:35 the next morning I found one of the Seronera males encrusted with blood, his left thigh ripped to the bone, and his body covered with punctures. Tatters of his yellow mane were strewn around the fallen leader. He breathed heavily.

At 6:55 a Masai pride male walked slowly to within five feet of

Vast stage for a survival drama of timely entrances and exits, the Serengeti Plain sustains the greatest concentration of hoofed animals on earth. The cast includes these thousands of wildebeests—but only one lion per four or five square miles. Wildebeests trek in seasonal migrations: to the treeless plain in rainy months, into acacia woodlands when the water holes dry up and grasses shrivel in the sun. Lions cannot live permanently on the plain; their base is the woodlands with its dependable resident prey, such as Cape buffaloes and impalas. Months may pass without prides seeing any wildebeests.
GEORGE B. SCHALLER

the vanquished one and, challenged only by a baleful glare and a low growl, ambled off. By 8:30 spasms racked the Seronera male's chest. His breathing grew erratic, then feeble. His bladder emptied. At 8:35 a muscle in his right thigh quivered and his pupils became very large. The king was dead.

Contrary to popular supposition, males have an important role in a pride. They are not just indolent parasites that leave the rearing of cubs and hunting to the lionesses. They provide security. Soon after this Seronera male died, males from three other prides moved in, drove off his male companion, and fatally mauled several cubs. The pride took nearly two years to regain its stability. In that time the death rate of its cubs was twice that of neighboring prides.

Lionesses in a typical pride do 80 to 90 percent of the hunting while the males seem to laze along. But a male is so conspicuous — his mane looks like a moving haystack when he attempts to stalk — that his participation would increase the chance that prey would detect the group. And the presence of males in the rear protects the cubs that as a rule lag behind stalking lionesses.

At a kill, when only a portion such as the forequarters remains, a male often appropriates it, keeping the food from the lionesses but sharing it with the cubs. This also guarantees meat for late-arriving cubs. Lionesses are not generous; even mothers may cuff their

starving cubs away from food. I once saw two cubs each snatch a gazelle's leg and flee from the kill. Two lionesses pursued them, bowled them over, and took the food.

The lion pride lacks a rigid hierarchy. That system, in which some members accept subordinate roles under the dominant members, governs many animal societies. Among lions, however, the weak fight for their rights. They may suffer when food is scarce; several cubs starved to death in the prides I studied. But a hierarchy would not help the weak, for it would most likely be based on size: The large males would eat first—and a male can devour 75 pounds of meat in a sitting. Then would come the lionesses, and finally the cubs. Such a system would lead to the dissolution of the pride, since animals low in the hierarchy would seldom get a meal.

I have seen a pride grow lean in its homeland while well-fed nomadic lions trekked behind migrating herds of wildebeests and zebras. What survival value does retention of a territory have? Intimate knowledge of the terrain no doubt increases hunting efficiency. I also noted that large cubs accompanied only 5 percent of the nomadic lionesses, compared to 23 percent of the pride lionesses. Though half of the cubs in a hungering pride may die, the stable, protective social environment holds down the death rate, increasing a cub's chance of survival. And such natural selection

Their cubs underfoot, lionesses walk the Serengeti in travel formation—single file, with some on the flanks. Returning from a night's hunt, they head for a shady riverside. When seeking prey, they will change formation, stalking on a broad front; cubs then keep to the rear but close enough to watch and learn techniques of the hunt. When they are a year or more, they may help pull down prey. By age 2 they may hunt alone.

GEORGE B. SCHALLER

makes for reproductive success—the continuity of the species.

Adapted to a regimen of feast or famine, lions may go for days without eating. With an assistant, William Holz, I followed one elderly nomadic male continuously for 21 days, keeping track of him on dark nights by means of a radio transmitter on a collar around his neck. He ate on only seven of the nights.

Dietary pragmatists, lions eat whatever meat they can find, including animals dead from disease and those scavenged from other predators. Lions are vulture-watchers, for they have learned that there is meat when these birds drop from the sky. At night, when hyenas punctuate a meal with their maniacal whoops and laughs, the lions run toward the sound. Often they find the remnants of a kill. On the plain, lions had scavenged about half of the carcasses on which I found them feeding. In the woodlands, where hyenas were not abundant, the figure was about 20 percent. Lions did not always win; on the plain at night they were driven away from their own kills by hyenas 44 percent of the time.

The lions' scavenging habits are not surprising; catching a meal is not that easy. There is a continual evolutionary race between predator and prey, a race with no winner. Constant predation, weeding out the stolid and the slow, produces alert and fleet prey.

Most prey avoid thickets, travel in single file, and are always poised for flight at water holes. Reedbuck may crouch when threatened; impalas flee in spectacular twisting leaps; topi can scan for danger from atop termite mounds. Lions must learn the preferences and idiosyncrasies of each species and apply this knowledge to the

In a superb display of strategy, four lionesses trap a herd of Thomson's gazelles. The quartet, after stalking together, splits. Two circle far around the herd, then wait (foreground), hiding until the other pair approaches. Sighting the hunters, the panicky prey scatter. But, boxed in, not all can escape. In this hunt witnessed by the author the lionesses rushed in from four sides and caught two gazelles.

DRAWING BY GEORGE FOUNDS

hunt. The lion usually seeks prey under cover of darkness. Hiding behind tall grasses and shrubs, it waits until an animal shows signs of inattentiveness or has lowered its head to graze. Only then can the lion steal close enough to risk a rush.

The effectiveness of predator-prey strategies is reflected in hunting success. I found that when lions stalked topi—an exceedingly alert species—only 14 percent of the attempts produced a kill. But when lions pursued the bumbling warthog, the kill score rose to 47 percent. The success rate for Thomson's gazelles was 26 percent and for wildebeests 32 percent. Two or more lions hunting together did about twice as well as a single stalker.

I once saw a young Cape buffalo bull battle a lioness for 90 minutes and lose only his tail, proving that a vigorous animal may fight off a solitary hunter. But against a hunting party cornered prey has little chance. When five male lions attacked an old buffalo bull, he managed to retreat into a swamp. The lions waited patiently on dry land. Suddenly choosing what I can only call suicide, he walked out toward the lions.

One grabbed his rump and another bit his shoulder. Slowly, methodically, they turned him on his back. The third and fourth lions moved in, one biting his throat, the other his muzzle. He suffocated in ten minutes while the fifth lion stood and watched.

Group living obviously has many benefits: division of labor, as when one lion guards a kill from vultures while the rest go off to slake their thirst; life insurance for the sick animal unable to hunt

Wheeling between life and death, three Thomson's gazelles escape –this time (above). But when a lioness singles out a Tommy at a water hole, there is no escape: he reacts a moment too late; she fells him with her hooked claws and kills him in the dust of the chase. A painted stork stands unruffled by the drama.

Lions devote little time to hunting. The author found that they average about 21 hours a day "sleeping, dozing, and staring into the distance." Nor do lions pay much heed to wind direction (though they catch gazelles three times more often stalking into the wind than with it). Short powerful limbs, padded paws with retractable claws, and long canine teeth make lions supremely adapted to stalk, overpower, and slay the quarry that for one fatal instant slackens vigil or speed.

HUGO VAN LAWICK

but allowed to subsist on the efforts of others; schooling for cubs that could not learn hunting techniques on their own.

Other species in the Serengeti achieve similar benefits by maintaining social systems modified to their own ways. Jackals live in pairs rather than groups. The two common species in the Serengeti —the black-backed jackal of the woodlands and the golden one of the plain—occasionally travel in small packs. But the pack consists of a mother, a father, and their grown pups. Jackals, once dismissed as mere scavengers, hunt down most of their meals. I have seen them kill gazelle fawns, hares, a newborn bat-eared fox, an infant vervet monkey, mice, and even beetles.

Hyenas live in clans, whose flexibility resembles the pride's. Deprecated for centuries as cowardly scavengers, hyenas have won new respect as hunters thanks to studies by Hans Kruuk, a founder of the Serengeti Research Institute (page 246).

The society of the African wild dog is founded on the pack, which, unlike the lion pride or the hyena clan, remains cohesive no matter what the quarry. The nomadic dogs, traveling in their packs of 5 to 30 members, spend a few days in one spot and then move on to a new area perhaps 25 miles away.

A 40-pound wild dog needs help pulling down, say, a 550-pound zebra. And cooperation pays off in a meal that satiates every member of the pack. When the hunters seek smaller prey, however, each dog must expend extra energy, for the pack keeps hunting until everyone is full. To take small game, then, solitary hunting would seem to be of greater advantage. But a nomadic species cannot afford to scatter; all members must stay in contact.

Each year, for about three months, a pack confines its roaming to the vicinity of a den where pups—as many as 16 in a litter—are raised with care strikingly different from that given lion cubs. A lioness always seems torn between self-indulgence and maternal devotion. In wild dog society, where males and females share equally in the care of the young, pups have priority.

Let me describe a typical hunt that begins from a den full of pups and their doting elders. Before setting out, the dogs exuberantly play and push their muzzles into the corners of each other's mouths —a greeting display derived from the food-begging behavior of pups. The activity probably gets the hunters in the right mood for a

Stalemate in the game of prey and predator: Giraffes, sable antelope, and zebra stand too far from the lioness for her to catch them—and they cannot risk drinking. Each knows precisely how close it can approach. Given a head start, prey animals, running at about 40 miles an hour, can escape a lion. But the odds are more even when the name of the lion's game is stalking.

communal endeavor—just as a pep rally does for us. One or two dogs stay at the den to guard the pups as the rest trot off, usually behind a lead dog. The most successful of the Serengeti predators, the dogs will undoubtedly find food; nearly 90 percent of their hunting expeditions result in kills.

Twelve dogs top a rise and see a gazelle herd ahead. Breaking into a run, several dogs chase after one part of the herd, several take off after another. When one female gazelle breaks away from the herd, a single dog runs after her at about 40 miles an hour. She flees in a wide arc, a typical prey maneuver. The pack shortcuts across the arc and two dogs dart in front of the gazelle. She swerves, only to face two other dogs. Trapped, she slows to a walk. In moments the pack tears her to pieces.

Crowding around a kill, dogs bolt their food but without the snapping and snarling characteristic of lions. They share the spoils. A lame dog was once seen limping to a kill after his packmates had eaten all the meat. The latecomer went around begging until one of the dogs regurgitated a piece of meat for him.

With a belly full of meat each hunter returns to the den and there throws up some of it for the pups as well as for the guards. When pups are old enough to tag along on a hunt, the food-sharing becomes even more remarkable. After a kill, the pups simply dash up and appropriate the whole carcass! The adults, no matter how hungry, step back and permit their young ones to gorge. The adults eat the remains—if any.

Scientific considerations aside, I must admit that the wild dogs' social system appeals to me more than the lions'. I recognize the irrelevance of comparing the dogs' food-sharing, cooperation, and devoted care of the young to our human values. Each animal society, including ours, represents an intricate adaptation to the environment. And each society has proved its success by its existence.

While watching predators hunt and kill, I felt sympathy for the prey. Yet I also felt a strange tie, an emotional kinship, with the predator. As anthropologist Sherwood L. Washburn noted, hunting "is a way of life. . . . In a very real sense our intellect, interests, emotions, and basic social life—all are evolutionary products of the success of the hunting adaptation."

Sometimes, as I walked alone across the plain with dry grass rustling beneath my feet and columns of wildebeests streaming under the hazy sky, a consciousness of the past pervaded me. To the east a faint dark line of acacia trees traced the course of Olduvai Gorge in whose depths Louis S. B. Leakey has unearthed remains of hominids, precursors of man. I would climb a kopje—a wind-worn granite outcrop jutting from the plain—and imagine myself as a small hominid, squinting against the glare, scanning the horizon for descending vultures, a sign of meat to scavenge. I wondered about the life of such a hominid, a primate by inheritance, a carnivore by avocation . . . existing as both hunted and hunter.

We naturally look to the monkeys and apes for clues to the social system of early hominids. But most nonhuman primates, though banded in groups, are basically vegetarians and confine themselves to relatively small ranges. Man, long a far-roaming hunter, has supplemented his diet with meat for at least two million years.

Since social systems are affected by ecological conditions, it seems reasonable to look at the animals that resemble early man ecologically—the social carnivores. The lion and the wild dog probably resemble early hominids in their social system more than do, say, the baboon and the gorilla. No nonhuman primate society typically shares food, hunts cooperatively, and has a division of labor.

By studying the carnivores, I believe we can better understand the spectrum of possibilities open to early man. The sophisticated cooperative hunting techniques used by the lion and wild dog suggest that early hominids had the ability to use game drives, lie in ambush, and encircle prey. It seems logical to assume that group size was flexible, adapted to the size of available prey.

Hominids had to fit into a certain ecological niche to reduce competition with other predators. In the two million years of the Pleistocene Epoch, when saber-toothed cats and giant hyenas prowled the savannas, man faced formidable competitors. Could he exist by merely scavenging? Or would he, like today's carnivores,

The king of beasts fights—and sometimes dies—in tests of his sovereignty. A monarch, whose luxuriant dark mane proclaims his maturity, evicts a young male that snarls defiance. Males are forced out of the pride if they do not leave by age 3½.

Mauled by three rivals, a king lies dying, tatters of his mane littering the grass around him. A lioness of his pride watches over him. He had been guarding a zebra kill and a lioness in heat when the males, from a neighboring pride, attacked. They fought by the lion's code: Pounce when odds are right.

"Poured out like honey in the sun," lions lounge in the shade, displaying the supple grace that inspired Anne Morrow Lindbergh's sweet simile. From Aesop to Born Free *the lion roams literature, a model of savage beauty. Studies of its life-style inspire behaviorist George Schaller to speculate whether similar forces shaped the societies of the lion—and man the hunter.*

both hunt and scavenge? My colleague Gordon Lowther and I decided to seek answers by simulating life as early hominids.

We camped on the banks of the Mbalageti River, whose pools lure thousands of zebras, wildebeests, impalas, buffaloes—and predators like us. We had not invented the spear or any other projectile weapon, but we could scavenge—walking along stream beds and in thickets looking for lion kills, watching the skies for circling vultures, scrutinizing each herd we encountered for sick prey. Several times we stumbled upon groups of lions. As soon as they saw us, they fled with a growl; tradition commanded them to avoid the most dangerous of all predators.

In about 125 miles of searching we found four freshly abandoned lion kills—three zebras and a wildebeest—with only bone marrow and brains left to eat. Our one lucky find lay beneath a tree laden with vultures: a buffalo bull, dead of disease.

We captured a sick zebra foal after a brief chase, then released it. And once we saw a young giraffe behaving abnormally as it followed its mother. Stalking closer, I saw it was blind. I grabbed its tail to simulate capture.

Adding up what we had "caught" in a week, we concluded that a small group of hominids could have lived in a similar area by capturing sick animals and scavenging. But, when the migratory animals moved on, the hominids would either have to move too or hunt healthy prey—unless they subsisted primarily as vegetarians.

Man is inextricably linked to his dual past, carrying within his frame both the terrible power of the predator and the frailty of the ape. Today, as man looks toward the future with primordial apprehension, he must learn to understand how the past has molded his life and will shape his destiny. What I sensed on the Serengeti, Carl Sandburg saw in his poet's eye:

There is a wolf in me . . . fangs pointed for tearing gashes
. . . a red tongue for raw meat . . . and the hot lapping
of blood—I keep this wolf because the wilderness gave it
to me and the wilderness will not let it go.

 .   .   .   .

There is a baboon in me . . . clambering-clawed . . . dog-
faced . . . yawping a galoot's hunger . . . hairy under
the armpits . . .   . . . ready to snarl and
kill . . . ready to sing and give milk . . . waiting—I
keep the baboon because the wilderness says so.

86

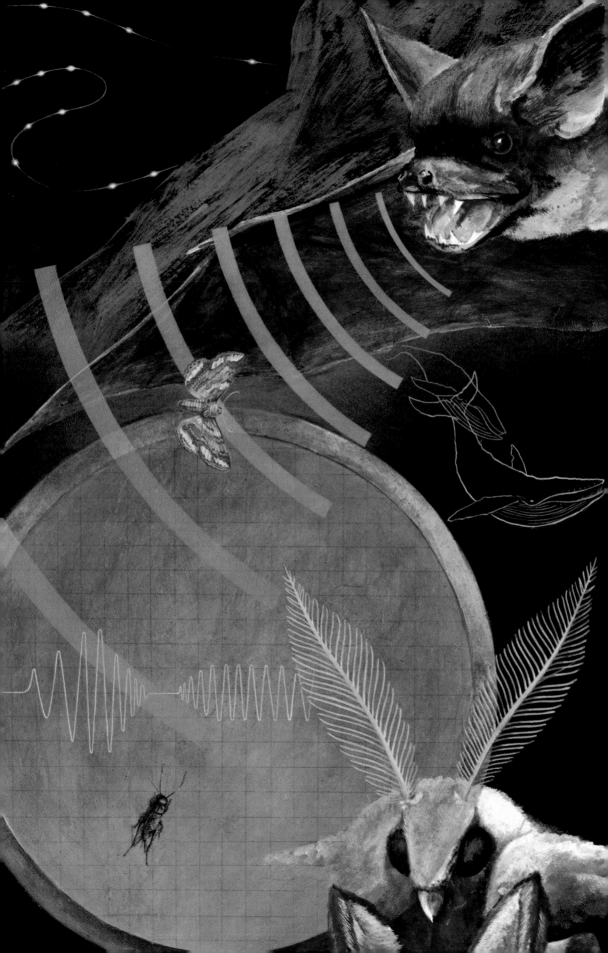

SAYING IT
WITH SIGHTS AND SOUNDS

Richard D. Alexander

Consider a summer evening in the country. What comes to mind? The flashing of fireflies, clatter of katydids, pulsing chirps of crickets? Perhaps the repetitious droning of a bullfrog, the booming dives of a nighthawk, the calls of a whip-poor-will, or the quaver of a screech owl? Maybe even a coyote's bark or the snort of a white-tailed deer. And faint odors and a freshness on the breeze. Sights, sounds, and smells, but for humans mostly sounds—sounds so characteristic that adding them as background to a dimly lighted movie or television scene gives the illusion of night.

Anyone who stays awake the entire night in a rural or wild locale will notice a hush as the night animals cease their activities just before dawn. Then birdsong bursts out of the grayness. When sunlight strikes across the meadow and into the forest undergrowth, a whole new bustle of activity begins. And again there are signals.

Anywhere in the world that men live—jungles, deserts, prairies, marshes, mountains, forests, or seashores—countless signals are broadcast. They are exchanged among the myriad creatures that have always lived with us, creatures helpful, harmful, or neutral to human existence. Primitive man's surroundings, moreover, must have resounded with many signals we will never experience. And because his life and happiness often depended upon them, he was probably more sensitive to animal signals and their meaning.

How many people today are aware of the messages in the behavior of animals they see at the zoo? A cheetah pauses beside a dead branch a few feet in the air, turns and shoots a jet of urine upward against it—a marking of territory under natural conditions. A lone howler monkey swings rapidly back and forth and emits long, ear-splitting cries; such howls in the jungle probably help monkeys keep acceptable distances between troops. A female rhinoceros assumes a stance that causes sexual arousal of the male. Zoo crowds often miss the special insight into the life of the animal which such signals afford.

Some signals are less easily overlooked. People usually stand in awestruck silence before the challenging stare of a big male gorilla. Once at a zoo in Miami I joined a group watching a new ape that crouched, terrified, in one corner of his exposed cage with his face buried in his hands. After a long silence a tiny, sympathetic voice from the crowd asked softly through the bars, "¿Qué pasa?—What's the matter?" A friend of mine was similarly impressed by the behavior of two chimpanzees in the London zoo separated by a wooden wall, solid except for one tiny hole. The chimps were sitting on either side, each gingerly poking a finger into the hole until fingertips touched; then both screamed with ecstasy or excitement or loneliness—or who knows what emotion.

Pulses of light and sound, subtle scents and touches link the lives of earth's social creatures.

PAINTING BY RICHARD SCHLECHT

My studies of crickets, katydids, and cicadas over the past two decades have taken me along many of the back roads of North America, Europe, Australia, and various Atlantic and Pacific islands. In rural areas, particularly those not yet invaded by roaring, whining motors, animal signals are more prominent than in cities, and people seem more aware of them. In a remote part of Mexico I once stopped to tape-record and collect a tiny cicada singing along a forest border. From a nearby hut two small boys approached timidly. I felt like a trespasser, and my headphones, 24-inch parabolic reflector, and other elaborate gear had to be a complete mystery to them. So I explained in halting Spanish that I was searching for a small *chicharra* singing in the grass. The boys nodded slightly but gave no other sign. I supposed they didn't really understand. But a few minutes later I heard in the distance, very faintly, a large forest cicada. When I lifted my head to listen, one boy murmured, *"Chicharra grande* — Big cicada."

What are animal signals all about? Why do they exist? How did they become complex and consistent and recognizable? Beyond mere curiosity, such questions interest us because communication is a central theme in the broader field of animal behavior. It refers

Sunrise shimmers a quiet pond,
gold-powders the white faces
of Queen Anne's lace, and cues
a company of daytime signalers.
An old field gone back to nature,
Edwin S. George Reserve serves
University of Michigan students
as a living biology laboratory.
In the vicinity of Crane Pond
55 species of crickets, katydids,
grasshoppers, and cicadas sing
at various times of day and night.

A nocturnal pair, snowy tree
crickets, meet on a twig.
The male at left has called in
the female by scraping the edge
of his left wing on the toothed
underside of the right. When she
touches his antennae, he turns
and offers fluid from a gland
on his back. Mating follows.
Songs differ with the species
and the phase of courtship. Only
male crickets sing, stepping up
the tempo on warm evenings.
The snowy has been called the
"thermometer cricket." The number
of chirps delivered in 15 seconds
plus 40 gives the approximate
temperature in degrees Fahrenheit.
Neighboring males synchronize
chirps into a steady, rhythmic
beat. Nathaniel Hawthorne
thus reviewed their nocturne:
"If moonlight could be heard
it would sound like that."

ANTHONY A. BOCCACCIO. BELOW: LOIS COX

91

STEPHEN COLLINS, PHOTO RESEARCHERS. ABOVE: LARRY WEST, FULL MOON STUDIO

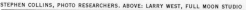

Twin voices of the night, signals of the tiny cricket frog and the true katydid, a cricket relative, may confuse the human ear but not others of their kind.

Passing air over vocal cords between lungs and a resonating throat sac, males of the inch-long amphibian call in concert from a breeding pond. In answer come females—and more males to amplify the chorus. Excited male frogs or toads may clasp any moving object of mate size; to cover errors they have evolved a signal for "release me."

The katydid, conspicuous on dogwood, makes sound as does the cricket but is left-winged. (The toothed file is under its left wing, scraper on the right.) As with crickets, two tympana, or membranes, on each foreleg just below the "knee" provide highly directional hearing. Named for its raucous call, the flightless treetop chorister rarely is seen on the ground.

92

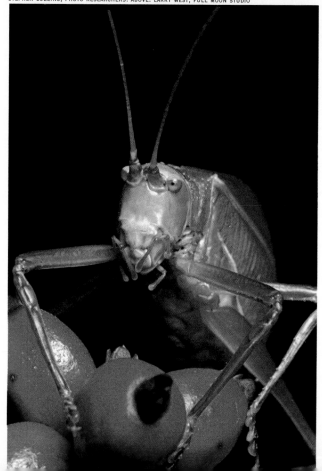

to systems of interaction by which animals transfer information — by which they effect changes in behavior. Communication opens the way to all the competitive and cooperative acts that almost define the word "animal."

The formal study of animal communication is one of the youngest disciplines in biology. Understandably so, for many signals are tuned to receptors so different from man's that he has missed their subtleties. The ultrasonic echoes used by bats, the intricate dances of bees, and the chemical messages of ants come readily to mind. But even the songs of some well-known birds are delivered too fast for human ears to appreciate fully the intricacy of their patterns and the nuances of their sounds.

Once I taped the song of an indigo bunting and played it for a group of biologists. Most of them recognized the song at its normal speed. Then I replayed the tape at one-quarter speed to show what they had missed. They were astonished. I had previously noticed that this particular bird ends his song with a little flourish that at the slower speed sounded just like the opening bars of reveille. As a joke I had whistled the remaining bars of the tune and spliced it onto the tape. At normal speed, no one had noticed the addition.

Even when human senses are adequate to pick up an animal's signal, its meaning may not be obvious. The response may be too subtle. Or the effect may not take place immediately. People, as we know, can absorb a message, store it, and act on it hours, weeks, even years later. But the slow-acting effects of some animal signals seem more remarkable. Sounds made by brooding parakeets or ring doves will set lone neighbors, out of sight, into the complex cycle of changes that accompanies breeding: nest-building, gonadal enlargement, and loss of feathers from the brood patch. The male and female of these tropical species depend on stimuli from one another to synchronize breeding behavior. Social insects such as termites and ants communicate by means of body chemicals called pheromones (from the Greek *pherein,* to carry, and *horman,* to excite). Pheromones produced by reproductive members of a termite colony can cause others genetically like them to develop into sterile workers or soldiers rather than reproductives.

Charles Darwin discussed the probable origins and biological function of familiar kinds of signals: cricket chirps, bird songs, canine postures, even human facial expressions. He noted that some seem to derive from "direct action of the nervous system," as with blushing in embarrassment or bristling with anger. Some derive from "serviceable associated habits" — lifting the eyebrows when surprised, he thought, might be a by-product of seeking a clearer view. Thirdly, a signal may convey a clearer message by being opposite to another of known meaning. A dog's submissive crouch, for example, is opposite to its aggressive rearing forward. Biologists still use Darwin's three generalizations in discussing some expressions and gestures of animals and man.

Darwin could only speculate about the functions of most signals; he was unable to analyze structurally any but the most obvious visual signals. Only in the past generation has technology given us the tools we need to make objective, repeatable studies of signal structure and function. High-speed portable motion picture cameras and ultrasensitive film capture rapid-fire visual signaling. Sound-recording equipment is sensitive well beyond the human hearing range. Gas chromatography and a variety of other techniques enable us to analyze and manufacture chemical signals.

Superior tools gave me an advantage over earlier biologists when I began to study North American field crickets. Earlier generations of entomologists had studied these crickets mostly in museums. They had examined, measured, and compared dead specimens. But because field crickets looked so much alike, even under the microscope, they were all believed to be of one species. It was known that cricket songs, produced by the adult males, differed widely. With portable high-fidelity sound equipment I was able to go into the field and bring back the songs along with the specimens. A laboratory device converted the taped sounds into audiospectrographs, or sound pictures. When these graphs were analyzed and the insects grouped according to their sounds, minute variations in body structure and size were found to follow the groupings. Further work showed that populations so identified do not interbreed in the wild. More than 20 species of North American field crickets have now been identified.

> *O cricket, who cheats me of my regrets, the soother of slumber,*
> *Muse of ploughed fields and self-formed imitation of the lyre,*
> *Chirrup me something pleasant. . . .*

The Greek poet Meleager in the first century B.C. thus praised the cricket's melodious calling. An ancient to us, the poet is modern compared to his subject. Fossil evidence indicates that these insects have been noisy since the heyday of the dinosaurs; body parts were adapted for making and hearing sound at least that long ago.

Evidence from comparative insect studies suggests that at an even earlier period crickets were simpler, cockroach-like creatures that did not signal acoustically but did shake their wings during courtship. An animal's means of signaling always derives from senses and body parts that originally served some other function. Female mosquitoes, for example, attract males by vibrating their wings. Birds signal with the colors and motion of their feathers, basically a body covering. Deathwatch beetles make a ticking sound by knocking their hard shells on wood. Various creatures use chemicals in body secretions as trail markers or beacons to find one another. Vocal animals, from frogs to man, make use of their breathing apparatus to produce sound.

As a particular way of communicating takes hold in an animal population, the body parts evolve along with the new function. Those animals that developed a complex social life developed many signals, often using more than one sensory channel.

Man himself is the super-communicator. He signals acoustically, visually, chemically, and by touch. To these "natural" kinds of signals his technology has added radio and X-ray waves and numerous others ways of passing and preserving information. He communicates on more different subjects than any other species. His complex social life leads to talk about sex, children and grandchildren, politics, social tolerance, war and peace. Not limited to the concrete or the close at hand, he discusses such abstractions as self-awareness and death-awareness and the nature of the universe—even communication itself. Unlike any other organism, man also communicates about events far removed in time and space, and he does so on a massive scale.

Man's communication is so complex that investigators searching for unique aspects of human behavior have increasingly focused on language. *(Continued on page 101)*

SENSES AT WORK

Dark-tipped feathers rim the heart-shaped sound collector that is a barn owl's face—all the better to hear with. Sensitive eyes enable the bird to see as well in starlight as man can under a full moon. But the night marauder also uses sound to take prey. And it hunts with consummate skill; in 25 minutes one night an owl was seen carrying 16 mice, 3 gophers, a rat, and a squirrel to its young. Densely packed, curving feathers form the collecting wall. Filmy overlying plumes— here partially clipped to reveal the channeling understructure—let sounds through. Ears beneath the downy veil have a hearing range less broad than man's, but can pick up many sound frequencies at fainter levels because of the efficiently shaped face.

95

Precision in pitch darkness

A telltale rustle in the blacked-out laboratory, and a barn owl arrows from its perch toward the unseen source; spread talons imprison the prey. What enables the owl to hunt so successfully when it cannot see? Its ears are proportionately larger and have a more specialized inner structure than most other birds'. Also, in many owl species ear openings are not placed symmetrically—presumably enabling the bird to sense direction in both horizontal and vertical planes at the same time. So discriminating is the barn owl's hearing that, after an alerting sound, it needs but one more to locate the source.

Leaving a perch, the owl pushes off head-forward on a flight path set by its ears. When there is light, the bird glides swiftly, feet tucked back. In total darkness it flaps its wings to slow its flight; feet swing pendulum fashion. At the last moment the bird turns end for end, talons replacing ears on the target line.

In another lab experiment a speaker emitting mouse noises was hidden under a sheet of paper. A switch on the perch stopped the sound the moment the owl took off, yet the bird regularly struck the covering. Talon holes always were carefully spaced to cage a mouse lengthwise (left); in the wild the owl maneuvers so the claw pattern aligns with the mouse's path. The silent-pinioned hunter makes mid-course flight corrections if it gets a mid-course clue.

DRAWING BY GEORGE FOUNDS

An eye for every need

One eye for the quarry, one for the foe? It's a matter of choice for Parson's chameleon (above). Its turreted eyes swivel separately. When an insect meal nears, the lizard's eyes coordinate while its long, sticky tongue whips the morsel in; then they diverge again.

Such oddball eyes make survival sense, having evolved to fit each owner's life-style. The stargazer, a bony fish that lies on the ocean floor, eyes upturned, luring food fish into its mouth, probably has no need of the detailed visual data essential to swift-moving predators like the cat or eagle. Scores of gemlike eyespots stud the mantle inside a scallop's fluted shell, but they sense light and motion only dimly; chemical sensors help the mollusk elude its nemesis, the starfish. The mosaic image perceived by the robber fly is also crude —despite the bulging bulk of its multifaceted eyes. Facets do not focus, yet compound eyes efficiently sense the movement of prey.

The keenest of eyes belong to birds of prey and sight-guided mammals. Flexible lenses focus sharp images from light rays entering an opening expandable for dim light. The cat's mirror-backed orb makes dual use of each stray glimmer. Feathery frame and bony hood shield an eagle's eye, among the most acute.

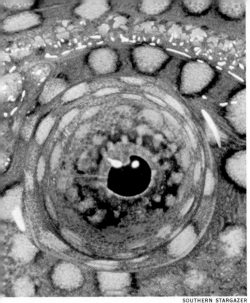

SOUTHERN STARGAZER   DOMESTIC CAT

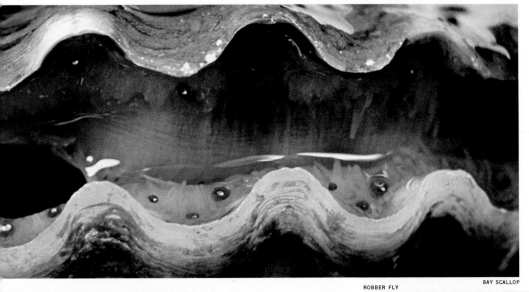

ROBBER FLY   BAY SCALLOP

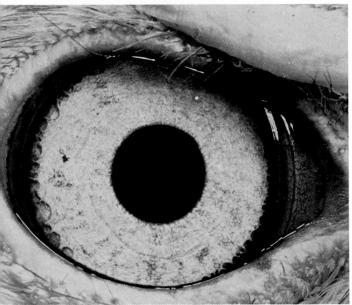

BALD EAGLE

A sense of direction

A cold-blooded denizen of the jungle favors warm-blooded prey. Special senses help find it, even in the dark. Looped on a limb, the South American emerald tree boa can detect a nearby bird or small mammal by means of heat receptors in shallow grooves along the lips. The paired right and left pits probably help the six-foot constrictor zero in on prey and strike with precision. Snakes lack external ears but are sensitive to vibrations. The forked tongue, too, is a sensor, flicking about, carrying airborne particles to a chemical analyzer, called Jacobson's organ, in the roof of the mouth.

Many creatures of murky waters rely on smell and hearing to find food, mates, or spawning places. Some fishes use electric organs to navigate or to stun prey. Another special sense that fishes possess, the lateral line system, consists of rows of pores in the skin leading to fluid-filled canals. These organs, sensitive to minute currents and vibrations, help fish swim in protective large schools.

Man, they say, is the only species with "true language." Despite its complexity the structure of human language can be reduced to fundamental building blocks comparable to those of animal signals. Language is composed of two major kinds of units. The most basic are *phonemes,* the smallest divisible units of sound. Phonemes often correspond to letters of the alphabet. The smallest units that convey meaning are groups of phonemes called *morphemes,* and they correspond roughly to words, phrases, and sentences. The smallest complete message can be observed to bring a particular response from others using that communicative system.

I can remember a feeling of considerable pride and confidence accompanying my initiation into a new signaling system. It was my first day of school in a one-room country schoolhouse in Illinois. Sometime in midmorning I urgently needed to visit the outhouse in the far corner of the schoolyard. But talking was prohibited, and so was leaving one's seat except to sharpen a pencil or get a book. I was at a loss and sat in misery. Then I saw a boy near me hold up two fingers and, following a nod from the teacher, depart the room. Instantly my two fingers also shot into the air. But the teacher shook her head, frowned, and said, "Not until Leonard returns." When Leonard did return I had already realized, by the length of his absence, the significance of two fingers. Confidently I held up one finger and took my brief leave, a full communicating member, as I saw it, of the new society.

Humans don't go around speaking phonemes to one another: They deliver whole messages. We can assume that animals don't spend much time giving incomplete signals either. The basic units of communication—the level at which function becomes apparent—are revealed by the ways animals combine and shuffle gestures, movements, sounds, or touches to make messages. If an animal always puts together what at first appear to be two different signals, never giving one without the other, we can be fairly sure that both parts are required to give the message accurately and completely.

The chirps of field crickets can be subdivided into sound pulses made by individual strokes of the wings. But playbacks of individual pulses do not cause other crickets to respond. Only certain minimal groupings and patterns of pulses make a complete message. As early as 1913 a German biologist, Johann Regen, sent the chirps of a male European field cricket through a telephone and observed that a female was attracted to the receiver. The experiment has been repeated with other species and refined with modern sound equipment. We now know also that female crickets respond only to the chirps and trills of males of their own species.

The only way a cricket can make his sounds is by stroking his wings back and forth. He produces a pulse by rubbing a row of teeth against a scraping edge, which together make up a "stridulatory" apparatus. More than 2,000 cricket species make sound this way; sometimes as many as 30 live in a single small field or forest. Only rarely, and only when they do not breed at the same time, do different species have the same pulse pattern. The patterns thus provide a rich signal grammar, or syntax. Comparative studies through analyses of recorded cricket songs reveal how messages are constructed and what are rough equivalents of phonemes and morphemes in human language.

Most of the examples used in biology to describe signal function come from insects. A major reason, I believe, is that higher animals with more complex behavior use many

Windborne odor lures a gypsy moth to a mate

11.1 mph **6.7** mph

different kinds of signals in quick succession or simultaneously. The functions of insects' signals are relatively easy to analyze.

Crickets are not only cheerful laboratory companions, they also make excellent subjects for study. They are easily reared in captivity. Different species can be hybridized (even though they rarely crossbreed in the wild). Surgery can easily be performed on them. Biologists have thus been able to study the development and neurophysiology of signals in crickets more thoroughly than in other animals. We know, for example, that 19 pairs of muscles help produce each pulse of cricket language. Those muscles are coordinated with a single impulse from a clump of nerve cells making up a pacemaker in the central nervous system. Other pacemakers modulate the stream of pulses into groups, or bursts, of pulses. These patterns, usually called chirps or trills, are the actual messages, comparable to morphemes. Each species has a distinctive repertoire, from one to six different signals.

Studies involving hybridization show that differences in the pacemaker rate which shape the distinctive species signals are inherited. A cricket gives the right chirps for his species the first time he tries. Unlike birds and mammals, most crickets hatch long after the previous generation has died. This means there is no "culture" involved in an individual's chirping—no learning from hearing a parent's song. Neither in a laboratory that is sometimes a cacophony of chirp and trill nor out in a world of even more alien noises has anyone yet been able to change a cricket's tune. There is much we do not know about how a cricket acquires his song, but we do know he does not learn it by listening.

How can we generalize the roles of animal signals?

Basically, the life function of every organism is reproduction. All its activities can be viewed as a means to that end. In order to live to reproductive age an animal must be able

2.2 mph

Written on the wind, the volatile scent of a female gypsy moth advertises her readiness to mate. A male (below) reads the message with his feathery antennae and moves upwind to find her. The potent alcoholic substance, called gyptol, attracted males from 2.8 miles away when wafted on a gentle wind of 2.2 miles an hour. Stronger winds scatter the odor, shortening and narrowing the effective area, shown by conic sections in model at left. In still air, some males detect odor but cannot find the source.

Gypsy moth larvae defoliated 1.4 million acres of American forests in one year. Decoy scents to keep males from finding females may offer a nontoxic way of curbing moth populations.

to get food, to avoid predators and disease, and to remain or go where the climate suits it. Then the animal must produce offspring that are able to do the same things. Some of these activities do not necessarily involve more than a single animal of the species. But in all sexual species, at least the last activity—production of offspring—takes two and therefore requires that male and female communicate.

Sexual activity begins with finding a suitable mate and producing a fertilized egg. In some cases it ends only after a long period of parental care of the offspring. Pair formation is therefore the most nearly universal context of cooperation and competition. And sex signals are the most prominent and diverse in the animal kingdom. They also appear often to form the basis of communication used in aggression, territoriality, and alarm.

Sex signals include not only the familiar bird songs, cricket calls, and firefly flashes but also the less obvious odors of a female moth or of a bitch in heat. Chemical signals, though perhaps the least perceived by man, are among the most ancient and the most widely used in the animal world. Odors of female moths can be enormously powerful stimuli. In one of the pioneer studies of sex signals in insects C. V. Riley more than 75 years ago released a marked male silkworm moth 1½ miles from a caged female. The male was on her cage the next morning. Recent work indicates that in a slow, steady wind the female's potent sex pheromone may recruit male moths at even greater range.

As one might predict, the signals responsible for forming pair bonds are never alike between species that live together, no matter how similar the species may be otherwise. Human beings rely upon whole constellations of signals of a visual, acoustical, and chemical nature, and form long-term pair bonds. It is thus not easy for us to understand the importance of distinctiveness in the sex signals of species that rely chiefly or entirely upon a single signal—an odor, a flash, or a chirp—and whose pair bonds may last only as long as a single copulation.

Some insect species are so much alike that biologists can find scarcely any way to distinguish them except by their dissimilar sex signals. Similar species may avoid mix-ups by signaling at different times of day or in different seasons. Or they may live in different habitats in the same geographic region. Those active in the same place at the same time evolve different signal structures.

Pair-forming signals operate in two ways. In the first system one sex, when ready to mate, stays in place and broadcasts a signal. The other sex cruises about, either homing in on an intensity gradient of the signal, or in the case of chemical signals sometimes simply moving upwind or upstream. In crickets, frogs, and most birds, the male takes a calling station and the female is the rover. In moths the female is a stationary chemical signaler and the male moves to her. Females of some moth species have lost their wings and legs and evolved into little more than bags of eggs and scent.

In the second major system both sexes signal. One begins with a long-range signal; the other responds. Thus they set up a reverberation of signals and responses and gradually move together. Fireflies and katydids start courtship this way.

The insect that stays in one place and emits a sound would seem especially likely to be eaten by a predator. It is not surprising that most insects signal only at night, when few insect-eating birds are active. Generally, those that venture daytime noises either call from burrow entrances, as with field crickets, or have excellent vision and flying ability, as with grasshoppers and cicadas.

In the cricket family, those that live in trees tend to sing only at night. The correlation is so close that in some sister species the one living in the grass sings much in the daytime, but its tree-dwelling relative almost never does. Bird predation is a fact of life for tree crickets, shown also by their close color match with the leaves and twigs of the host trees. The pine-tree cricket of eastern North America not only has a reddish head and green body to match the sheath and needles of pine foliage, but when disturbed it positions itself so it looks almost exactly like a pine needle.

Whatever an animal's means of signaling, it seldom finds a clear channel. The world is full of competing sights, noises, and smells. As a result, some animals have evolved ways of sending signals that carry great distances, can be continued for hours, and are essentially flawless from a physical point of view. The enormous larynx of a howler monkey, the hollow abdomen of a male cicada, and the balloonlike vocal sacs and larynx of a frog all act as loudspeakers to increase the power of the emitted signals.

"All's fair in love" seems to be true even at the insect level. Take the "false katydids." During an evening of sexual activity a male flies from one singing spot to another, stopping briefly to signal. When a female answers, he alternates signals with her and slowly flies and walks toward her. He holds his forelegs, bearing his auditory organs, in odd positions. He slowly tilts and waves them, listening directionally.

John Spooner, who worked out how these katydids form pairs, gave me an impressive demonstration one night in my backyard in Michigan. For a while he answered a signaling male by striking the blade of a pocketknife against a baby-food jar. Then he told me to turn on my flashlight and look around. I was amazed to see half a dozen other katydid males forming an arc in front of us, slowly waving their forelegs and moving silently in

EDWARD S. ROSS

Parental ploy: A killdeer performs a "broken-wing" display
that may save its nest from plunder. Predators, seeking
the easy mark, have evolved the ability to detect
when birds are hurt. Nesting killdeers exploit the exploiter.
The robin-size plover lays its eggs on open ground.
Male and female share incubation duty. When an intruder
comes near, the nesting bird, perhaps alerted by the
alarm call of its mate, hurries away. It flops about,
beats the dust, and, dragging a wing as if it were broken,
leads the interloper directly away from the nest. When
the predator closes in for the kill, the faker flies away.
If the brood has hatched, the parental maneuver—
a step removed from an alarm signal—may save lives
by causing the young to freeze or squat out of sight.

Some animals employ an opposite kind of mimicry,
pretending to be armed when actually helpless. Certain
tenebrionid beetles can spray a noxious chemical to keep
from being eaten. A similar-looking beetle lacking scent
glands fools some predators by mimicking the headstand
position its armed relative assumes when about to spray.

Wet and woebegone trio of month-old coyote pups—survivors of a flooded den—and an adult that has found winter food celebrate two of many howling occasions. Members of the genus Canis command a rich repertoire of social whimpers, barks, and growls as well as the long plaintive howl reminiscent of a prairie party line. Said one writer who found music in the calls of wolves: "Like a community sing, a howl is . . . a happy social occasion. Wolves love a howl. . . . troop together, fur to fur."

our direction. Evidently they were creeping toward the responding "female" whose location was exposed by the one male alternating calls with her.

It may be to the female's advantage to broadcast her responsiveness to every male in the vicinity and let the "best" win. It is not clear how a male fights such competition. But in alternating with a female he does throw in an odd little noise much like that of the female, so quickly after her signal that it seems a part of her sound. This little confuser perhaps gives some trouble to the males freeloading on the signaler.

A different kind of opportunism confronts some firefly species. James Lloyd has discovered that some females of the genus *Photuris* are capable of answering the flash code of males of the smaller *Photinus*. When boy meets girl, however, the *Photuris* females seize the responding males and eat them.

Long-range mating signals probably evolved from close-range courtship. Courtship signals are usually softer, less distinctive, and more erratically delivered than long-range pair-forming signals. When two tree crickets are courting at close range, the male raises his forewings, exposing a chemical area on his back which attracts the female into mating position. Vibration of some ancestral male's body as he lifted his wings in courtship was probably the evolutionary forerunner of stridulation.

The signals that reproductive individuals or pairs use to secure and hold territories are also associated with sex. The functions of a territory differ with various animals. In crickets the territory is the area from which a male excludes other males which might compete with him for mates. Females have little stake in such a territory.

In birds, on the other hand, male and female commonly cooperate in a monogamous unit to raise as many healthy babies per clutch of eggs as they can. In many songbirds, each parent must make dozens of insect-capturing trips each day. Keeping other birds out of the nesting area forestalls attacks on the young and also saves the nearby food for the nesters' use. Although surprisingly little is known of the significance of the males' songs for their females, the songs are known to be aggressive signals between males.

Just past midnight one summer several years ago, I was driving through Shiloh National Military Park in southern Tennessee. On an impulse I stopped among the shadows of its great oak trees. The park was bathed in moonlight so bright I could read the inscriptions on the pedestals telling of the events of April 1862, when Union and Confederate forces fought here. It was easy, standing alone in the moonlight, to envision the battle scene. Off to my left I could hear two nightjars—a northern whip-poor-will and a southern chuck-will's-widow—posted very near one another and calling in a furious and rapid antiphony. As I read from the plaques and glanced across the slopes reconstructing the battle, I was suddenly chilled by a tremendous boom and rumbling roll of sound across the woodland to my right. Seconds passed before my mind came back to the present and accepted the noise as a sonic boom from a jet plane now fading into the night sky.

As I climbed back into my car, though, the thing that kept ringing through my mind was the symbolism in the aggressive interaction of those two birds, Yankee and Rebel, that happened to overlap ranges here. So easy for soldiers on night duty to borrow their signals, I thought. And I wondered whether some innocent, feathered tops-of-posts hadn't been blasted into eternity during those few fateful days a hundred years before.

Man discusses, debates, argues—and worse. Animals also wield signals in competing for commodities having reproductive value: food and water, shelter, mates, and territory. For many animals, sex signaling ends with copulation. Other species care for their babies: organisms as diverse as burying beetles, honeybees, salamanders, toads, snakes, and almost all birds and mammals. Individuals in these species have more survivors by using part of their reproductive effort to reduce mortality of their young.

Some of the shift in effort is almost certain to involve the development of special communication between parental animals and their offspring. Such signals are constantly before us: the exchanges of whinnies and bellows and bleats and grunts by mares and colts, cows and calves, ewes and lambs, sows and pigs. The signals help parents and young stay together or find each other if separated. In some large social groupings, they help parents to distinguish their own offspring.

Until a few years ago it was generally believed that orphaned birds in large colonies were fed by other adults, and even that adults fed young indiscriminately, not knowing their own. In each case studied carefully, however, parents were shown to recognize their offspring by its voice or by sight and sound together. Obviously the parent who refuses to feed any but its own young will have a reproductive advantage. Herdsmen have long had trouble getting ewes or cows to adopt orphans. Sheep raisers sometimes pull the skin of a dead lamb over an orphaned one to get the bereft mother to adopt it.

Forewarned is forearmed in animal as well as human society. An individual that detects another's reaction to danger has more time to flee or ready its defense. An animal giving an alarm signal in some cases also increases its own safety. A bird sounding an alarm call may cause the flock suddenly to take to the air or scatter, so confusing the predator that it does not attack. A young baboon that screams when it sees a leopard may attract adult male baboons that can successfully bluff or repel the predator. But usually the individual giving a danger signal attracts the predator's attention; the alarm-giver takes a risk. Such behavior is often thought of as altruistic; that is, the signaler endangers himself in order to save his fellows.

How can natural selection favor such behavior? For the "selfish" individual that does not endanger itself for others seemingly would be safer, would produce more descendants, and eventually would be the only kind of individual left in the population. But when parents signal to protect their offspring, they may increase their reproduction over parents that do not. Even though a parent is sometimes killed by the predator, its signal may save a number of offspring. Similarly, a signaler may be preserving his own genes by saving his pregnant or brooding mate. An individual helping close relatives could feasibly benefit in terms of saving his own genetic material. Biologists call such partiality to relatives "kin selection." In humans we call it nepotism.

Alarm signaling is not confined to the breeding season nor to groups known to be composed of close relatives. In some cases, however, there is reason to believe the signaling is an outgrowth from parents protecting babies. A parent-offspring signal that is useful most of the year may, on rare occasions, be given in disadvantageous situations because greater precision in its use has not yet evolved. Even for the alarm-giver, a well-alerted flock is a safer place to be than a poorly alerted one.

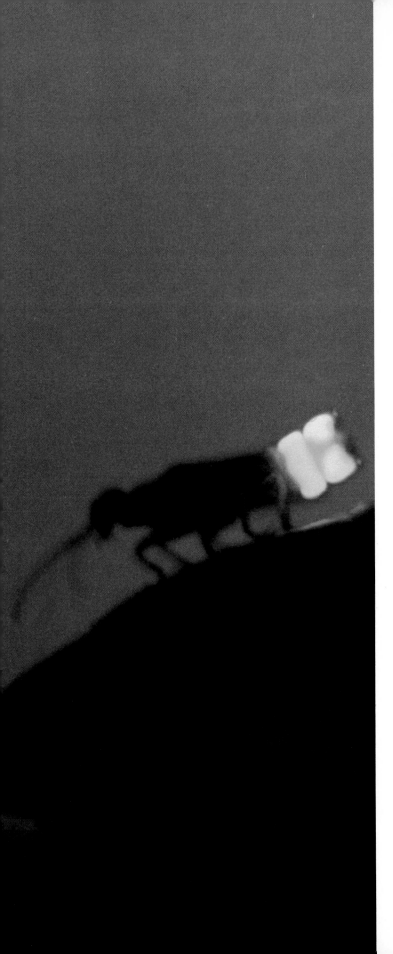

Built-in beacon gives the firefly
a visual signaling system
that works in the dark. While
most night creatures call on other
sensory channels—chiefly sound
and smell—luminous beetles
of the family Lampyridae blink
their abdominal lanterns.
A compound, luciferin, interacts
with an enzyme, luciferase,
and oxygen to make cold light.
Larvae as well as adults can
emit a glow, but most fireflies
studied flash only during pair
formation or when disturbed.
Flight paths of flashing males
vary; some species can be
identified by their J-shaped
swoops or periodic waggling.

Beaded trails of light (above)
show paths and rhythmic flashing
of five species in the American
genus Photinus, as they might
appear in a time exposure.
A male begins the signaling. When
a female turns on at the proper
interval, he flies toward her,
exchanging signals periodically.
Attraction is based entirely
on the flash code; females of
various Photinus species put in
airtight glass cages attracted
males as in the open.

Males of some Southeast Asian
fireflies, including Pteroptyx
malaccae (left), synchronize
flashes. Pooling their luminosity
may increase their ability
to attract females.

IVAN POLUNIN. ABOVE: VIRGINIA BAZA,
GEOGRAPHIC ART DIVISION

Most alarm signals probably begin when simple fright behavior of one animal is used by another to avoid the same danger. Once such an action takes on signal value, and it also helps the actor in some way, it may spread through the population.

Animals of different species sometimes benefit from one another's alarm signals. Baboons and impalas, for example, often forage together. Baboons have keen eyesight; impalas have keen senses of smell and hearing. Thus each can profit from the alarm signals of the other. When two animals gain from each other's presence, they develop not only social tolerance but also better ways of communication in succeeding generations.

Such mutual aid, or "reciprocal altruism," is prominent among animals that interact in social groups. Some animals recognize one another individually and are thus able to identify and discriminate against cheaters that do not act for the benefit of others when it is their turn. Some of the complex social communication of human beings seems to fit such a model. Trying to explain alarm signals in terms of natural selection is a topic of great current interest to biologists. Different theories have been offered, and important questions remain unanswered. But alarm signals are sure to figure prominently in the future study of animal communication.

By coincidence the peculiar beeping signal transmitted back to earth from one of man's first orbiting satellites sounded remarkably like an intriguing and seldom-heard animal communicative signal: the piping of a new queen bee still in her pupal cell. The piping coincides with the departure of the old queen and some of her workers to a new site. Listening to the one signal and thinking of the other, I was struck with how much man, the super-communicator among all organisms, has learned about signals and their meanings and how much he has yet to discover.

Satellite beeps and honeybee piping have more than their sound in common. Each signifies a beginning: for man an era of direct exploration and dispersal into unknown regions of the universe; for the honeybees a new era in the life of the colony.

The beeps and piping are each products of an enormous amount of social cooperation. And each is inextricably linked to that complement of cooperation, social competition. No satellite is launched without the power and coordination generated by thousands of people working together. No queen honeybee is produced without the cooperation of thousands of bees. But we can legitimately wonder how long it would have taken a single nation without a competitor to make those initial probes of space. And no biologist doubts that it has been the constant reproductive competition between honeybee colonies that has produced their complex, almost bizarre social existence.

Competition results in change, whether good or bad in the long run. In a social organism cooperation is a vehicle of that change. These two interwoven facets of behavior are the key to the fascinating and complex patterns of interaction in the animal world. Communication, because it is the sole means by which social competition and cooperation are carried out, is the essence of social existence.

Galaxy of fireflies sets a mangrove tree aglow. Here in tropical Malaysia swarms of Pteroptyx
*perform each night. A single male begins flashing; gradually others pick up the rhythm until nearly
all are blinking in unison. Responding females emit dimmer, unsynchronized flash patterns.
Signaling mechanisms that benefit an animal pass on to new generations, refueling the torch of life.*

IVAN POLUNIN

Jack W. Bradbury

THE SILENT SYMPHONY:
TUNING IN ON THE BAT

The tropical rain forest . . . a world packed with life . . . a world not so much of sights as of sounds. Giant trees reach 150 feet into the air to spread their crowns in a canopy of vegetation. Vines, ferns, lianas, and other growths weave a dense matrix between the forest floor and the covering. In this thick greenery no broad vistas open. Seldom can your eye find a line of sight greater than 30 feet. Even the illumination filtering down forms a patchwork of light and dark that tends to break up outlines. Little wonder it is difficult to see the animals of the forest—or for them to see each other. Little wonder, too, that sounds are so important.

Sound waves bend around trees and vines. They carry long distances while the animal that makes them remains hidden. At night, when it becomes so dark you cannot see your hand before your face,

Hanging "swing-swang to . . . tops of high trees"—as a scientist with Capt. James Cook in 1777 put it—giant fruit bats begin to stir at dusk on a Pacific isle. Wingspreads of as much as five feet make raven-size "flying foxes" the world's largest bats. The smallest, bamboo bats, are no bigger than the end of a man's thumb.

Asleep by day, bats dominate night skies. They evolved eons ago, perhaps from earthbound mammals, reaping the advantage of fewer predators and less competition for food during nocturnal hours. Fruit-eaters see at night with especially sensitive eyes, but most bats use echoes to navigate and hunt—making man's sonar crude by comparison.
GORDON W. GAHAN

sounds of the forest come into their own. And the champion among creatures that put sound to use is the bat.

Bats constitute the Chiroptera, the second most numerous order of mammals. (Rodents rank first.) The majority of bats are tropical. By day they roost and are often difficult to find; at night the tropic skies are filled with them.

In the dense lowland forests of Gabon in West Africa I have spent summers studying the sounds and social habits of bats. Often I paddled a dugout canoe on the Ivindo River and waited while the sun set. At that time of day the river becomes a major highway for ibises, hornbills, turacos, and parrots—all returning from foraging grounds to night roosts. Their calls and screams echo up and down the river; male monkeys in the trees add their barks and woofs to the din. Then, about half an hour before dark, the birds and monkeys settle down and the bats take over.

On the Ivindo, the first evidence of bats is a soft ticking sound. If you look carefully, you will see several large black shapes flying through the dusk at high speed and making sudden swoops and dodges. These are Pel's bats, whose 27-inch wingspreads make them the largest insectivorous bats in Africa. Soon other dark shadows join them, hawking over the river, streaking above the forest canopy, flitting around the larger trees.

All these bats use their own sounds to navigate and to locate prey. Most such vocalizations lie above the range of human hearing. But if we could hear them, we would be overwhelmed by a wild cacophony of chirps, honks, metallic notes, buzzes, and clickings. Bats form one of the most vocal of mammalian groups, and in the tropics their sounds reach a clamorous height.

If we happen to be at the right spot, we may be startled just before dark by an enormous bat with a moose-like head and a three-foot wingspread. It emerges from the forest and flies directly and rapidly to an island in the river. There it lands in the dense vegetation and begins a loud, metallic honking—audible to the human ear—that continues at a steady tempo of 60 to 100 times a minute. Other bats join the first, and the honking chorus begins to sound like a crowded frog pond. This is the fruit-eating hammerheaded bat, surely one of the most unusual of mammals.

Heads of males are swollen and grotesque in shape. Internally, the bat is equally bizarre. It has, relative to its weight, one of the largest larynges known, a voluminous and bony box. From that box comes its steady honk. Nightly during the dry season male hammerheaded bats congregate in traditional sites and emit their raucous honking to attract females. Each male in such an assembly appears

Warty lips, bulbous nose, and flaring ridges give hammerheaded males a face to titillate female bats. As many as 130 males may gather in leks—communal display areas known also among grouse and some hummingbirds. Each bat hangs head down (this photograph is inverted) and beats partly opened wings at double time to accompanying honks from a larynx which fills one fifth of his body cavity. Early in the breeding season the strident singing to attract females is heard for an hour or so after dark and before dawn; later it becomes a chorus continuous except for a midnight break.

Author Bradbury studied the hammerheaded bat through the cooperation of the government of Gabon and that of France, which maintains a scientific station in the equatorial land.

For bat studies at night the author used infrared TV systems and light-amplifying telescopes developed for the U. S. Army. High-frequency microphones and recorders captured bat sounds.
JACK W. BRADBURY

to have his own singing tree. Should another male fly too close, snarls and hoarse gasps meet him, and the occupant may lash out with partly folded wings. Females arrive after dark and fly silently along the array of males. Aroused, the males step up their honking. If a female finds a particular male interesting, she may hover before him or dart into the adjacent foliage. The male then responds with a staccato honking followed by a number of long buzzes. Several times the female may fly away and return. At last she lands beside him and mating takes place. Within minutes she leaves and the male resumes his honking.

Communal display areas of males for mating, called leks, are also found among certain birds. But among terrestrial mammals only hammerheaded bats and an antelope, the Uganda kob, use them. Where leks exist among birds, males usually have evolved showy plumage and displays as a result of competition for females. This also may explain the male hammerhead's grotesque facial features.

The hammerhead belongs to a suborder of bats labeled Megachiroptera. This group includes the familiar fruit bats of Asia and the Pacific islands. Megachiropterans have large eyes, two claws on the edge of each wing, and—except for the tomb bat—do not use sounds for navigating. Instead, their sensitive eyes enable them to see on all but the very darkest of nights. Most bats, including all the species in the United States, belong to another suborder, the Microchiroptera. They have only one claw on each wing, small eyes, and do use sound when flying. As the name suggests, microchiropterans tend to be smaller than megachiropterans. They also are wider spread, living on every continent except Antarctica.

In Trinidad I have studied one of the more fascinating members of this group, the whitelined bat. These little creatures—three U. S. dimes could balance one on a scale—live on the sides of large trees, often completely exposed to view. When roosting, they do not huddle, as do many bats, but space themselves so that no bat is closer to another than one or two inches. A colony, totaling as many as 40 bats, typically consists of males and their harems—each male with his own aggressively guarded territory.

Each night the whole colony leaves the roost tree to feed. In the early dawn the males return and start to squabble over territory. Two males face off along a boundary line and "bark" at each other. If barking doesn't settle the issue, they begin an elaborate "gland shaking." The gland is a small pocket in the wing membrane between the bat's shoulder and wrist. In a display the bat will thrust his wing forward, snap open the gland, and shake the wing several times. Presumably this disperses some glandular scent near the

BRUCE DALE, NATIONAL GEOGRAPHIC PHOTOGRAPHER

On reaching wings, a whitelined
bat hovers before a female
he has enticed to his tree-trunk
territory; he'll build a harem
of as many as nine for the day's
repose. The pair's cries take an
unvarying pattern: his a series of
chirps, hers a drawn-out rasping.
The tiny bats get their name from
dorsal marks that help camouflage
them on the bark of trees, such as
these silk-cottons the author
climbed (opposite) in Trinidad
to study whitelined behavior.

   Many bats roost in clusters,
but the whitelined keep strict
distances of an inch or two,
a boundary violated only by very
young babies. Highlights on the
upraised wing of the hovering
male reveal the gland he snaps
open in territorial squabbles
or in his courting displays.

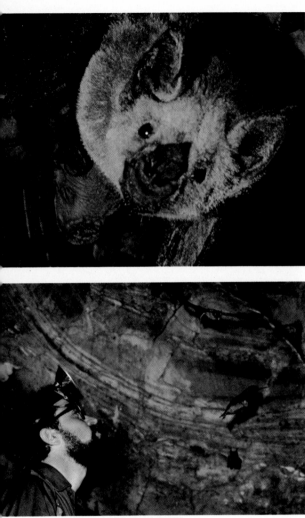

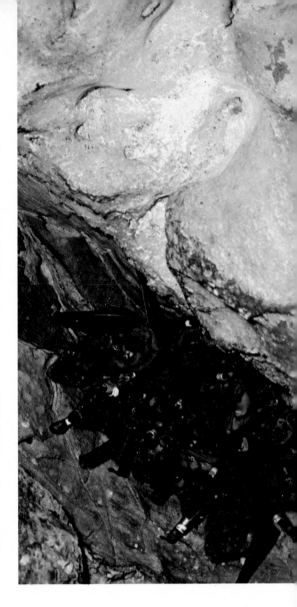

BRUCE DALE, NATIONAL GEOGRAPHIC PHOTOGRAPHER

Pollen-daubed face of a greater spearnosed bat betrays its raid on a flowering tabebuia tree in Trinidad. It eats insects as well as fruit—and even bats of other species sharing caves where it roosts. Its cave-ceiling clusters (opposite) may be all-male or have 10 to 100 females with a harem-master. Pink shapes are baby bats. Banding—right wing on males, left on females—aided the author (above) in observing behavior.

Oversize, mobile ears help bats pick up faint echoes from sonar cries. Noses with leaf-like tips mark bats that beam sound through nostrils; other bats use trumpet-shaped mouths.

other bat. Gland fights may last five to ten minutes, the males taking turns shaking and snapping. The bats may even lunge or fly at each other, but real battles are rare.

When females return to the roost, males break off their fighting and start to sing. To the human ear the songs are structureless twitterings. But recorded and analyzed electronically, the songs reveal their complexity. Each song has many elements: chirps, pure notes, raspy buzzes. The elements group into themes, much the way that many bird songs do. Different whitelined males sometimes include identical themes in their twitterings.

Females that land in a singing male's territory set off a singular hover display. The male flies from the tree in a shallow arc, curving back toward the new arrival. On beating wings he hovers in front of her, and both emit cries. He also may snap his gland.

Far from passive, the females interrupt their vocalizations to evict from the harems females they do not like. Thus while males are

hovering, singing, and shaking glands to retain females, the latter are chasing each other out. After about an hour of furious display and reshuffling, harems form for the day, and the whole colony settles down to groom and to sleep. Although whitelined bats have only a single annual peak of breeding, they perform displays and maintain harems all through the year.

Harems are no rarity among microchiropterans. In Trinidad caves I have seen this behavior in the greater spearnosed bat. These large omnivorous bats form dense clusters in cracks and potholes of cave ceilings; colonies may number in the thousands. Aggressive when in control of a harem, a greater spearnosed male rushes toward an intruding rival with beating wings and high-frequency vocalizations. Males without harems are amiable and often, after feeding, play together like puppies.

Another harem builder, the lesser spearnosed bat, has an elaborate vocal repertoire important in maintaining colony structures,

keeping groups together when feeding, and locating lost young. But it is in sonar (from *s*ound *n*avigation and *r*anging) that the cries of these and other bats reach amazing perfection.

The principle of bat sonar is simple: The bat emits a short pulse of sound and listens for an echo from nearby objects. Bats, however, have had millions of years to perfect this system. More than a mere detection of targets, it serves as eyes in the dark. In many species the sonar is so sophisticated that by "reading" the echo the bat can tell where the object is, how big it is, whether it is moving and how fast, and even its shape and texture.

Like the human larynx, the bat's vocal apparatus sets the air vibrating, thus producing sound. The faster the vibration, the higher the pitch of the sound. A pulse of sound in some bats' sonar consists of vibrations of a constant frequency—a single note. More commonly among bats the sound pulses shift from high-frequency vibrations to low. The human ear can detect sounds of from 20 vibrations, or cycles, per second to about 18,000. Most bat cries are of a higher frequency, thus inaudible to man. Typical sonar sounds of the little brown bat, for example, sweep from 90,000 cycles per second down to about 45,000.

Cruising bats produce only a few pulses each second. But as they near their targets, they step up the pulse rate to get more echo information. Faster rates mean individual pulses must be shorter, dropping to as brief as 1/2000th of a second. Physiological problems of producing such short sounds are considerable, yet bats do it thousands of times every night of their lives.

How can a bat extract complicated information from the echoes?

Time elapsing between emission of the pulse and return of the echo is one measure of target distance. Differences in intensity of echoes detected by the bat's mobile ears become a measure of direction. But distance and direction aren't enough. Physics tells us that best echoes result when sound waves are of a length equal to, or less than, the object which reflects them. Thus for a very small insect, the optimum would be a very small — short — wavelength. Now it so happens that the higher the frequency of sound, the shorter its wavelength; hence the bats' sweeping frequency range can cover a variety of object sizes. And since a target's shape and texture reflect some frequencies more strongly than others, the bat can "read" these varying echoes for the information it needs.

Air impedes the travel of sound; the higher the frequency, the greater the intensity required to propagate it. The energy in cries three feet from a bat's mouth equals that of a jackhammer pounding half a block away. Bats' ears have developed marvelous cushioning and muscle structures that shield their hearing from the effect of outgoing pulses — yet permit reception of echoes 2,000 times fainter than the emitted cry. Consider, too, that the bat must sift echoes of its own pulses from those of its companions, and the wonder of its sonar becomes truly staggering.

Emitting an outgoing pulse at the moment an echo from a previous one returns could obscure information about targets. Short pulses prevent such overlaps. Some bats, however, deliberately use overlap to intercept a target. The European horseshoe bat emits a long pulse of constant frequency. Vibrations from such sounds seem closer together when the source of the sound moves toward

Mirrored grace masks a moment
of drama: the thrust of a fishing
bat's claws to impale a swordtail
in a pond at the Bronx Zoo.
The bat's prey can be discerned
as a streak of orange at the
water's surface. In a twinkling the
fish will be crunched and carried
away in the bat's strong jaws.

Fishing bats skim fresh and salt
waters of Central America and the
Caribbean, feeding on minnows
detected by an exquisite sonar.
Narrow wings of 18-inch span
give the 2½-ounce bat ability
to lift a catch; its feet evolved
into sturdy, gaff-like talons.

Other bats have long tongues
to lap nectar or broad molars
to chew fruit. Insect-eaters
may spread a between-the-legs
apron to scoop large bugs on
the wing, raising them to sharp
fangs for shredding in flight.
Vampires have legs that give them
remarkable ability to leap and
walk about in stealthy approach
to quarry; they numb victims'
skins by licking, then bite with
keen incisors and lap the blood.

Vampires of the New World—
none exist in Europe, despite
folk tradition—and other bats
sometimes harbor rabies.
As with other rabid wild
animals, however, they seldom
aggressively attack man.

BRUCE DALE, NATIONAL GEOGRAPHIC PHOTOGRAPHER

a listener than when it moves away; a train whistle seems to rise, then fall, in pitch as the train approaches, passes, and goes on. Horseshoe bats take advantage of this "Doppler shift" in echoes. By detecting the slight shift in the apparent frequency, the bat can adjust its sonar, and then its flight, to home in on an insect.

Atmospheric conditions affect the transmission of sound pulses. Warm, moist air tends to filter out higher frequencies before lower ones; the latter travel farther, though they result in fainter echoes. For slow-flying tropical bats this poses no problems. But for the fast-flying Pel's bat, high-frequency sound couldn't reach out far enough; the bat would be on its target before it had time to react. Thus Pel's bats depend on longer-range, low-frequency sounds—so low they fall within range of human ears. The ticking when they fly is one of the few sonar pulses we can hear.

In each part of the globe bats have evolved amazing sonar mechanisms which make the night air theirs; they can hunt almost any kind of food which birds eat in the day. Some bats seek insects in the air, others pluck them off vegetation, and still others catch them on the ground. Bats gaff fish, feed on fruit or pollen, drink nectar from night-blooming flowers. Some vampire species drain small amounts of blood from large mammals, others feed on sleeping birds. Finally there are big carnivores which fly through the tropical forests at night and capture roosting birds or, perhaps by listening to their sonar, other bats.

Sound for bats is a way of life. Among megachiropterans, it serves social needs alone; the male hammerhead commits a large part of his body to producing a sound whose sole function is to attract females. Among microchiropterans, their phenomenal range of sound serves both social and flight needs. The whitelined, for example, calls females at 6,000 cycles per second, hovers with chirps at 12,000, barks at other males at 24,000, and hunts at 42,000.

Unfairly maligned by horror movie and eerie tale, the bat is a superbly adapted animal. It has few superiors as an insectivore; some species consume half their weight in a night—as many as 5,000 gnat-size insects an hour. Even the vampire must be accorded a place for what it truly is, a timid creature whose brushes with man are rare. Only in the last 30 years have we begun to understand the bat and its ways. Field and laboratory work have given but a taste of the enormously diverse use which these animals of the night make of their acoustic world.

It is a world as complex as the visual one, but hard to imagine. In part, this is what makes its study so fascinating; every new finding is unexpected. From bats we still have much to learn.

Leathery wings cloak a drowsy flying fox in a group startled by the camera's bright flash. Enwrapping while roosting helps megachiropteran bats regulate body temperatures. Microchiropteran bats rarely cover themselves, but often huddle in tight clusters. Males and females of many tropical species live together all year, but most temperate-zone bats segregate by sexes at birth time. Colonies of 20 million mothers plus young have been observed in one Texas cave.

For hanging upside down, bats have toes and knees that bend opposite to man's, and blood-flow systems whose valves work in reverse to ours. Auditory centers in the brain are exceptionally well developed.

In elaborate daily ritual, bats hang by one foot and use the other to comb and preen. Grooming with lips and tongue keeps wings moist and supple. Far from being the unlovely "flying mice" that tangle people's hair—as popular notion brands them—bats are unique among living mammals and valuable subjects for scientific study.

BRUCE DALE, NATIONAL GEOGRAPHIC PHOTOGRAPHER

Caryl P. Haskins

CONSIDER THE ANT
SOLDIER, BUILDER, FARMER

"Go to the ant, thou sluggard; consider her ways, and be wise: Which having no guide, overseer, or ruler, Provideth her meat in the summer, and gathereth her food in the harvest." So does the Bible's sixth chapter of Proverbs record Solomon's musing over the society of the industrious ant. Half a world away in pre-Columbian Mexico, there arose a legend, rendered into Spanish after the Conquest, about a talented worker ant that planted a seed from which all mankind was to develop.

Thus man's fascination with the ways of ants has extended over at least 29 centuries of his intellectual history. But the paths of men and ants had crossed long before then. Who knows for how many tens of centuries the ancestors of Australian aboriginals gathered the swollen workers of the deserts' honey ants to add a specially delicate dessert to a meager meal? Or for how long the peoples of Africa gathered around termite mounds to capture the giant queens of the thief ant, which, roasted in coals, provided a surpassing treat? With the building of man's earliest fixed shelters in Africa, with the collection of food stores about the first tribal villages, fierce driver ants must have forced themselves on the attention of the tribesmen. Dense columns of ants must have flowed like living rivers through every nook and cranny of hut and possessions, causing temporary evacuation by the human owners — but neatly clearing the premises of all vermin!

In modern times men of science have considered the ant, paying special attention to the suggestive parallels often found between the social lives of vertebrates — particularly of man himself — and the communal existence of ants.

As a general rule, ants live in a cooperative society of wingless female workers, winged males and winged virgin females, and a queen that regularly lays quantities of eggs. The workers build, repair and defend the nest, gather food, and nurture the eggs, nursing

Portrait of a worker: Australian bulldog ant, guarding a larva, displays the bulging eyes that bespeak outdoor habits, antennae for probing, and spiky mandibles for digging and biting. Its ferocious sting makes even scientific observers leery of a close approach.

The inch-long ground dweller, a lone hunter for insect fare, belongs to one of the most socially archaic of ant groups; communication among such ancient species may be limited to primitive exchanges of food. Workers feed the larvae, soliciting in return substances exuded through the larvae's skins. More highly developed ants use chemical messengers to mark trails and identify their nests, and as alarm and attack signals.

For science, "living fossils" like the bulldog ants offer a priceless opportunity to study at firsthand the evolution of social behavior.

PAUL A. ZAHL, NATIONAL GEOGRAPHIC STAFF

Biting and stinging, an African driver ant (genus Dorylus) *seizes a grasshopper which will be torn into bite-size bits. Like* Eciton *army ants of the New World, the drivers travel the tropics in raiding columns of tens of thousands, sometimes 1,000 feet long, that attack every living animal—even man. Indians clamp wounds with the huge jaws of* Eciton *soldiers.*

On the move, some army ants build a new bivouac every night. They entwine their bodies into a shelter for the queen and the larvae, which they carry from nest to nest, like the Eciton *nomads opposite.*

the brood through larval and pupal stages. Adaptable and numerous, the ants do all of this in an intriguing variety of ways.

Consider the legionary ants, such as the driver ant of Africa or the army ant of the New World; they have been called the Huns and Tartars of the insect world. Some species have workers that are blind or nearly so, yet they are extremely wide ranging. They live in communities of hundreds of thousands but are committed to perpetual wandering and raiding, some changing the location of their nests each night. Remarkably enough, the nests may be formed by the living bodies of the ants themselves; in some South American species such nests hang from trees like swarms of bees. From these masses raiding columns issue, searching every stick, stone, and blade of grass for live insect prey to feed the colony.

Amid all the bustle and haste and seemingly total confusion when a new bivouac is established, amid all the struggle with the environ-

ment, the immense extended community manages to retain its coordinated activity, its organized wholeness.

How can this be done? We are perhaps nearer to an answer today than we have ever been. Two important factors are the chemistry of ant communication and the role played by the queen.

An extraordinarily delicate sense of smell and a sense of touch are the ant's most important mediators of behavior. What these senses convey to an ant may well be combined in ways utterly unfamiliar to us. For on the ant's mobile antennae the sensory elements for smell seem intimately mixed with organs of touch.

Ants produce substances that act as chemical messengers and cause reactions in other ants. These substances, called pheromones, or "social hormones," help guide ants along trails and provide them with a colony odor, a kind of uniform that identifies nestmates. Through its combined chemical and tactile—or topochemical—sense the ant can distinguish as aliens not only other species but members of other colonies of its own species.

Because it is increasingly evident that pheromones are critical to the structuring of ant societies, the study of the production and distribution of these substances has become one of the most exciting pioneer areas in the field of social insect behavior.

Among the most important of the pheromone glands are the paired mandibular glands in the head and the Dufour's gland in the abdomen. In the genus *Formica*, which includes slave species as well as species that enslave them, we encounter one of the most dramatic illustrations of the role these glands have played in the evolution of ant behavior. In less developed *Formica* ants, the Dufour's gland is small and its chemical products induce panic and

129

flight among members of a colony. The first workers aware of danger spray the substance from the tips of their abdomens; the result is similar to the warning call in a flock of birds. More highly developed *Formica* species have a larger Dufour's gland that contains additional chemicals. These create attraction and aggression rather than panic. So the ants in this group tend to congregate at points of emission, biting their victims and spraying yet more chemicals.

This development culminates among the slave makers of *Formica sanguinea*. When a column of these ants pours into a nest of a slave species, the secretions from the raiders help focus the attacks in regions of highest concentration of vapor—right where the victims and their young are likely to be most dense. But among the victims the spray acts as an alarm pheromone and their reaction, predictably, is to scatter, abandoning their young to the marauders.

No leadership can be identified in an ant community comparable, say, to that in a pack of wolves. Yet there is nothing in all the vertebrate world that compares to the total social influence of the queens in some species of ants. By their very presence, by the pheromones they secrete, and by their egg production, these queens organize, regulate, and maintain the colony.

A striking illustration of this is the function of the queen of certain New World legionary species in determining the raiding cycle. At the height of foraging activity the columns raid until evening and, under cover of darkness, establish a bivouac at a new site. This nomadic phase lasts 16 to 18 days. Then the colony enters a statary, or sedentary, phase, dwelling in less transitory nests for 20 or 21 days. The queen's abdomen rapidly attains huge proportions as eggs develop. Meanwhile, all the larvae in the colony have spun their cocoons and now, as pupae, lie inert. With the need to feed the larvae removed, raiding slows down.

About a week after the start of the statary period, the queen begins to lay eggs at a tremendous rate—40,000 or more in a few days. In the days that follow, the eggs hatch into tiny larvae. But from the standpoint of the colony an even more important event occurs. The pupae of the previous brood all hatch together in a mass emergence of hungry young adults, stimulating the older workers. Raiding picks up sharply. Columns take up the trails once more, now augmented by clumsy, inexperienced newcomers getting their

Living bridge of army ants forms a shortcut for raiders. Triggered by the light of dawn, the troop spills from its bivouac along trails chemically blazed by ants in the van and reinforced by each follower. Sometimes a group accidentally isolated starts circling and locks itself into a "suicide" mill— which stops only when the exhausted ants march themselves to death.
MERVIN W. LARSON

131

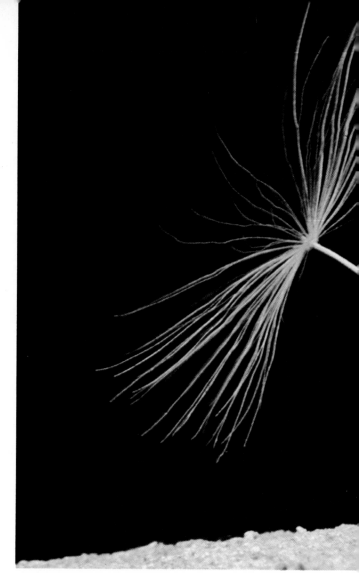

Bringing in the sheaves,
a foraging harvester ant of the
southeastern United States
carries home a dandelion seed.
At the nest workers will hull it
and store it in an underground
gallery like the granaries
exposed below, tended by
worker ants of different size.
Pogonomyrmex badius *is the
only North American member of
its genus whose workers are
polymorphic, or of varied forms.*

To prevent germination, ants
set damp seeds in the sun
or bite off root portions. They
dump chaff on an outdoor
rubbish heap, along with debris
and dead colony members.

Foragers find their way safely
to the nest among densely
grouped—and hostile—colonies
by chemical trails, visual
landmarks, or orientation to
the sun. Experimenters have
found that when they erase
a trail this harvester can home
by sunlight even if the sun
is obscured. As with bees,
the eyes of ants are sensitive
to a zone in the ultraviolet
region of the light spectrum
that humans cannot see.

ROSS E. HUTCHINS

first practice at following the tumbling river of ants. Shortly comes the emigration to a new nest. Insect booty flows into the colony at an ever-accelerating rate. Nourished by it, the larvae grow rapidly, thus inciting the workers to ever more spectacular raids as the colony moves through its nomadic phase. The larvae mature, exerting their maximum stimulative effect. And then, suddenly, once again a whole generation spins its cocoons. Once again, as the brood stimulus is withdrawn, the colony enters its sedentary phase. Once again the queen prepares for a fresh round of egg laying, and the whole process begins anew.

Harvester ants follow a way of life as unlike that of the legionary ants as can be imagined. Far from being nomads, they are firmly tied to one nest site excavated deep in the ground with an extensive and complex system of galleries or chambers. Often mounds of dirt mark the entrances. From them, workers issue forth, sometimes in files, to gather the seeds of grasses and weeds. Here

again, the column does not have any obvious leaders. Yet by the time it forms, all the ants seem to know where they are going, perhaps from familiarity with the terrain gained by simple observation. In some species, scouts may mark the trail with scent by dragging their abdomens on the ground.

Most impressive of all agricultural ants are the fungus growers, which not only harvest but actually cultivate their crop in underground gardens. Abounding in the tropics of the New World, they may defoliate all the trees within their reach, for they climb trees and plants to cut sections from the leaves. Carried to the nest, the leaves are chewed and deposited in the garden beds, contained in chambers sometimes as large as a football. The leaves nourish fungus, and the whole colony feeds on the fungus sprouts.

Like harvesters and fungus growers, the honey ants look to the future. They inhabit dry country and endure long periods when water, and food itself, are hard to find. They, too, hoard supplies—

Parasol, or leaf-cutting, ants of the genus Atta parade with leaves for their fungus gardens, their only source of food. They clip sections from plants (upper left) and carry them down tunnels into subterranean cities teeming with thousands of workers, including stay-at-home soldiers. In garden chambers the ants mince the leaves, deposit a drop of fecal liquid on each piece, and place them on the bed as fertilizer.

A gardener of another New World genus, Trachymyrmex, adds a leaf (upper right) to a fungus bed abloom with whitish, bulbous "kohlrabi heads," on which the ants feed.

A young queen on her founding flight carries a bit of fungus, cultivating it with her feces until she rears workers to take over her new colony's farm.

PAUL A. ZAHL, NATIONAL GEOGRAPHIC STAFF
UPPER: RUDOLF FREUND, PHOTO RESEARCHERS,
AND (RIGHT) ROSS E. HUTCHINS

but within the bodies of colony members. These "repletes" beg so much liquid food from returning foragers that they swell into balls, sometimes barely able to move.

The workers of all higher ants use their crop—the "social stomach"—to accumulate liquid food for long periods and then distribute it as needed. The size of the crop and the complexity of the valve mechanism that seals off its contents from the digestive system vary greatly, being more highly developed in more advanced groups. And this seems logical, for the sharing of food throughout the colony, together with the pheromones that may be mixed with the food, lies at the heart of ant social organization.

There are ants that tend other insects just as men tend cattle, milking them for honeydew. The insects, commonly aphids, or plant lice, feed on the sap of plants and exude clear, sugary liquids. The ants may keep them in herds, transporting them from place to place on the host plant as stems dry up.

It is a short step from the keeping of other insects by ants for their usefulness to the keeping of ants by other ants for the same reason. Here a behavioral series can be traced. It includes members of the genus *Formica,* denizens of woods and fields and often of mountaintops in the Northern Hemisphere—and perhaps the most behaviorally complex of all ants.

The various stages in the series may not be directly related in evolution, for we have no fossil evidence of ancestral patterns to guide us. But the sequence represents a logical progression of true slave keeping within this related group of ants.

We may begin with the wood ant of Europe, an exceedingly

Gargantuan in their nest, honeypots look like berries in a human hand.
Denizens of arid habitats where food supplies fluctuate, honey ants gather
nectar in times of plenty and pass it on to workers called repletes.
Bellies swollen into transparent spheres, these living storage tanks hang
immobile in special chambers, regurgitating the food when hard times hit.

MERVIN W. LARSON. LEFT: ROSS E. HUTCHINS

active red-and-black insect whose populous colonies live in huge thatched mounds. Long ago *Formica rufa* discovered a rich source of animal food in the larvae and pupae of a less aggressive species of *Formica*. In concerted raids it carries off larvae and cocoons and often stores them alive in chambers of the nest. Many are later eaten and more must prematurely die. A few attain adulthood. These individuals, knowing no other environment, settle down in their new colony and may serve as helpers there.

Other species representing the next stage of progression also carry out raids and return with the young, but allow most of them to hatch. As their numbers grow with each raid, the slaves come to form a distinct caste, aiding their captors in housekeeping duties, particularly as nurses looking after the young, allowing their captors more time and energy to conduct further raids.

In the third stage the relationship of slave and slaveholder is almost reversed. This development is well illustrated by some species of the genus *Polyergus*. Workers of these slave makers are warriors with pointed mandibles capable of piercing a victim's brain. Home from a raid, they turn the stolen larvae and pupae over to the slaves already in the nest, taking no further interest in the care and feeding of the brood. Indeed, even if the innate behavior patterns governing such care had not been lost to them in evolution, *Polyergus* ants would still be physically incapable of performing these duties. Their mandibles are so specialized for military functions that they lack the teeth essential for handling the growing brood. Incompetent in any activity except raiding, *Polyergus* workers must even depend on slaves to put food in their mouths.

Young queens of slave-making species are typically unable to found colonies in the manner of their independent relatives, which excavate a cell and rear the first brood in isolation. Instead, after the slave-making queen has flown and mated, she may seek to return home. Or she may search out a weak community of her slave species. Entering the nest, she assassinates the foundress queen and most of the adult daughters, and appropriates the pupal brood which is nearing maturity and so requires a minimum of care. As these young workers hatch, they adopt their imposter mother and rear her eggs to maturity, beginning a new colony of slave makers.

How did social life rise among the ants? We can only speculate, for no truly solitary ants, fossil or living, are known to give us a clue. Recently impressions of two ants were discovered in fragments of the Magothy amber of New Jersey, reckoned to be about 80 million years old. Their anatomy suggests that they were workers, and that their species, therefore, was already social.

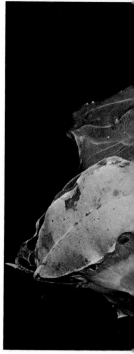

Weaver ants high in a tree
team up to pull the edges of
two leaves together. Then,
in one of the most complex
behavior patterns known among
social insects, other workers
of this arboreal species carry
half-grown larvae down the seam,
using them as silk-producing
shuttles to "sew" the leaves in
place and make a nest (left).

Inducing the larva to exude
silk from a duct near the mouth,
probably through some means of
stimulation, a shuttle-bearer
carries it back and forth,
crisscrossing the strands into
a formidable web. All larval
silk goes into building the nest,
so when the time comes
for the larva to pupate,
it spins no cocoon. Yet it stays
safe and dry in the silk-bound
chambers it helped build.

139

Dairymaids of the insect world—
like this honey ant tending
aphids on a vine in Connecticut—
keep insect "cows." The ants
milk their charges by stroking
them, stimulating the excretion
of drops of honeydew which the
ants lap up. Less sophisticated
herders wait until the "cattle"
voluntarily yield their sugary
secretion. Some ants move their
herds—which can include
mealybugs and leafhoppers—
from place to place as food
for them runs low; others
corral herds below ground
where they can feed on roots.
Some herders erect earthen
shelters for their cattle,
while others chew up fibers
and make paper sheds.

Another form of social
parasitism finds ants playing
host to guests that produce
desired secretions. One British
species carries caterpillars
to its nest and places them
among the ant larvae—which
the caterpillars devour
at leisure, while the ants
feast on caterpillar sweets.
In time the caterpillars
pupate and emerge in the ant
nest, then as butterflies
crawl to the surface, spread
their wings, and fly away.

J. A. L. COOKE

The typical ant society, in contrast to our own communities, is basically a family. It consists of a queen mother and the extensive population of her female descendants. Her offspring include wingless worker females, winged males, and winged females which will leave the nest as virgin queens. Like honeybee drones (page 54), the males do no work and are soon lost on mating flights with young queens setting out to establish new colonies. The queen's nuptial fertilization will last her a lifetime—perhaps 15 years. The average worker lives only a few years; males die soon after the mating flight.

Ordinarily the wingless female workers remain unfertilized. Yet they may lay eggs, and these will develop—usually into males, which do not contribute importantly to the colony's life. Thus the workers have little genetic effect on the course of ant evolution.

This specialization of labor, centering the significant reproductive capacity of the colony in the queen caste, means that since the colony became established millions of years ago it has evolved

much like a biological unit. The worker caste has become so specialized that it is incapable of maintaining an independent existence. And it has even become fragmented into subcastes.

Among the most highly developed fungus growers, there exists a whole spectrum of worker forms, ranging from aggressive soldiers almost as large as the queen down to tiny minims. These minute workers probably never leave the nest except to ride on the leafcutters or the foliage, their weight apparently not felt by the bearers. Their presence there was long thought to be accidental. It now appears the minims may repel the attacks of parasitic flies.

An interesting example of caste specialization occurs among some of the thief ants, which dwell in the nest walls of their larger hosts. They live by pilfering the forage brought in, or by stealing and devouring the hosts' young. Such species characteristically have tiny workers, well adapted to slipping unperceived on their nefarious missions, and huge queens, prized in Africa as a gastronomic delicacy. One of these giant queens may equal in bulk a thousand of her minute workers. So enormous is she that when she flies from her home colony on her mating journey she carries with her half a dozen workers clinging to her legs, their mandibles clamped to the hairs, which they just about equal in size. Later, when she has excavated her brood chamber, they may dismount, care for her eggs, and rear her first brood. Her own huge limbs and mouthparts are seemingly far too gross for the task.

Soldiers of the genus *Colobopsis,* whose colonies dwell in the pithy interiors of plants, often have heads adapted as plugs that can block their narrow nest entrances. Such soldiers may stand for long periods, their heads blocking the passage perfectly. When a worker approaches from either direction, a tap on the head or abdomen will prompt the soldier to stand aside.

It would be hard to imagine an organism more different from ourselves than the ant. At the level of bodily structure, at the level of sensory and neural endowment (10 billion neurons in the brain of man, perhaps half a million in that of the ant), the gap is tremendous. And as with individual ants, so with their social organization. The societies of ants and men are separated by an immense gulf. Ants are creatures of an age different from our own, indeed of another world, for their most formative period of evolution came eons before our remotest simian ancestors were born.

And yet, with all of this, we are constantly confronted by the extraordinary parallels found in our two societies—the hunters and the harvesters and the farmers. Are these parallels coincidence? I do not believe they are. There are too many and they are too intricate and often too suggestive to be coincidental. The answer may lie deeper and we are only beginning to explore what it may be. We know that many animals, unrelated and often quite different in inner structure but which have lived long in similar environments, have evolved to resemble one another to an extraordinary degree. The sharks and sturgeons, for example, can claim no immediate relationship and yet how alike they appear.

So it may be with the highly developed societies of ants and men. Perhaps living and evolving in similar social environments, subject to similar kinds of evolutionary pressures, subject to similar requirements of conduct that keep societies stable and permit them to grow, they and we, on more than a few occasions, have followed similar channels and discovered similar opportunities.

It is hard to think of a more rewarding task than that of garnering insights into these resemblances, to discover, perhaps, still-hidden facets of behavior in ants—and in men.

A red raider of the slave-making genus Polyergus *challenges a* Formica *ant (opposite)—one of the species commonly seen on our sidewalks—for her cocoon. If the defender strongly resists, the invader's sickle-shaped mandibles can easily dispatch her with a bite through the brain. Then the victor picks up its booty in the manner of the larva-toting robber above, and heads home.*

As ant slavery evolved, the masters became ever more dependent on slaves; Polyergus *cannot even feed itself. Some species exist as solitary parasites, perhaps well on the road to extinction. This, notes author Haskins, is "the great price . . . paid for the surrendering of independence."*
EDWARD S. ROSS

Roger Payne

THE SONG OF THE WHALE

Sunlight dancing on the ceaseless waves . . . myriad schools of fish darting just beneath like showers of silver. Such is our usual vision of the ocean. But this encompasses only the thinnest, uppermost layer. We give little thought to the underlying abyss, far greater in bulk, perpetually dark and very cold.

In the depths beneath the surface film, life consists mainly of tiny fishes adapted to the cold and the immense pressure. Because they live in darkness they must have trouble finding each other. The twinkling spots of luminescense sprinkled along their sides and bellies may improve their chances; as they move in shoals they look like slowly wheeling galaxies of stars.

Good visibility in seawater is seldom as much as a hundred yards; animals that rely on sight to find others of their kind are probably

Living archipelago of humpback whales erupts off Alaska during a herring feast. Breaching may stun many fish, making for easier pickings. Gulls fishing in the troubled waters can flap too close to an open maw — some have been found in whales' baleen, the hornlike slats in the upper jaw that filter out food.

Lauded by Herman Melville as "the most gamesome . . . of all the whales, making more gay foam and white water," the humpback today stirs new wonder with its wild, mystifying songs.

CHARLES JURASZ

145

able to meet only when they chance upon each other at close range. Those that depend on sound are far better off; sound travels much farther in the sea than light does.

If you lower a hydrophone into deep ocean, you may hear amid the abyssal murmuring the eerie mingled calls of some distant animal. At first scientists were baffled by these sounds, especially when they realized the faintest noises were remnants of extremely loud calls that had traveled a long way, weaving between the bright, undulating ocean ceiling and the oozy floor. In recent years we have found that many of the loudest sounds are the calls of whales. Some are very beautiful to human ears, surprisingly so since they have had a completely independent evolution — in a world as different from ours as a separate planet. Now, after millions of years of isolation, we and whales in a sense have come together; with underwater microphones we can listen in on their world.

I have spent many a day listening to whales through hydrophones and countless hours playing tapes of whale sounds. Of all I have heard, the most remarkable are those of the playful humpback. In both the Northern and Southern Hemispheres groups of humpbacks make long migrations, at times near continental shores, summering around the 60-degree latitudes. There these baleen whales sieve from the ocean unbelievably rich summer blooms of plankton.

In the south almost no land interrupts the plankton belt. Once it attracted earth's greatest concentration of whales. It also attracted whale hunters, and in this century they have all but destroyed the humpback there. Today the four small northern humpback populations probably outnumber their southern kin.

Humpbacks from one or more of the northern groups breed around Bermuda and the Antilles, and each spring for five years my wife Katy and I have gone to Bermuda to study their hauntingly beautiful sounds. With Scott McVay of Princeton, a dedicated whale conservationist, I analyzed the tapes and discovered that the whales were singing! The songs lasted from 6 to 30 minutes— far longer than the usual few seconds' duration of a bird's song.

Many biologists define a song as a complex, repeated pattern of sounds; in contrast the very short, repeated sounds of insects and frogs are more properly termed "calls." The definition does not imply a pattern pleasing to us, though in the humpback's case it is. Others have heard in these songs the sounds of lowing, moaning, wild shrieking, and wailing, but I have never found words adequate to describe them. Though we may find eventually that other whales also sing, the only singers definitely known in the mammal world

Setting a trap for whale sounds, Roger Payne anchors a sonobuoy off a remote Argentine coast— a winter haven for right whales. The floating rig picks up and broadcasts underwater sounds.

Right whales are rarely seen, much less heard. Slow swimmers that float when harpooned, they all but disappeared under pressure of 19th-century whaling. Here, in reef-fringed shallows, a remnant of the species comes to mate, calve, and play.

Eavesdropping on the hidden world of Leviathan is a chancy game; animals that may measure 55 feet and weigh as many tons can accidentally break the sonobuoy lines and gear.

WILLIAM R. CURTSINGER

are man, the whitelined bat, the bearded seal, and the humpback.

Why do humpbacks sing? We have no idea. They sing more at night and, it seems, on the wintering ground. But we do not know the singers' age or sex, whether they sing alone, in pairs, or in groups. One fact intrigues us. The high-pitched, high-frequency sounds are soft and varied, the low sounds loud and less varied. Why?

Given equal loudness, low frequencies go much farther through the ocean than high ones. Thus we would expect the highest humpback sounds to travel at most a few hundred yards before being lost in the background noise of waves; the lowest will go many miles. Unless we assume the whales are wasting a lot of time and energy on low sounds, the songs apparently contain information for two audiences—one nearby, the other at a distance. Or a song could be serving as a beacon that gives the listener some sense of distance, depending on how much is heard.

If the humpback is voicing its high-frequency repertoire for listeners close by, then this variety of sounds may be carrying a large amount of information. And our wonder grows at what it all means. The low frequencies would be expected to convey a simple message to a distant whale—perhaps no more than "There is a humpback here." The distortions of distance would smear out any subtle information; far away one would anticipate no more than a monotonous and repetitive signal, like the droning of a foghorn.

We need to know more about the loudness of humpback song to estimate range. But the sounds of another cosmopolitan baleen whale, the finback, are simpler and have been much studied. We can calculate roughly how far they carry. The result is surprising.

Let us lower a hydrophone into the abyss in the Pacific, hundreds of miles from the nearest shipping lane. The phone picks up a faint deep tone, its frequency around 20 hertz (cycles per second). It comes on and off like clockwork—in one typical pattern a one-second tone, then alternately 12 or 15 seconds of silence. This is

To the sea's ageless symphony the voices of whales bring a beguiling chorus. Scientists have picked up sounds from every whale species studied extensively—though none yet heard can match the eerie music of humpbacks. Poring over spectrograph charts—on which the humpback's rising whoops, whistles, bellows, grunts, and descending roars look like this—

Dr. Payne and Scott McVay found complex sequences that comprise the fabric of song. And each humpback sang its own variations on the species' themes.

The songs were widely heard on a recording, "Songs of the Humpback Whale." The strange, haunting cries turned men's minds to the plight of whales and inspired a musical piece for orchestra and whale soloists.

The whales do most of their singing while submerged. Lacking vocal cords, they probably make sounds with the larynx and by internally shunting air.

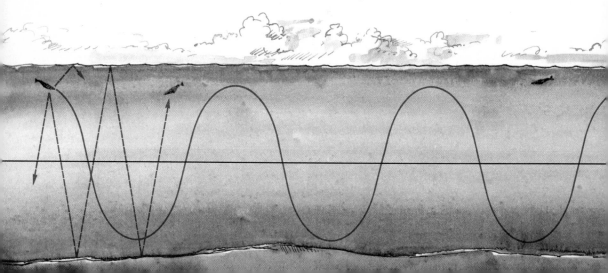

a call of the finback. We do not know whether finbacks are the only source of the sound—only that they are one of the producers of some of the sounds heard around this frequency.

At times 20-hertz sounds are very loud. When acoustics experts engaged in naval research first heard it, they did not believe an animal could generate so loud a sound. Scientists finally were able to identify one of the "20-cycles-per-second monsters" as a finback and to pick up some of these calls 100 miles away! But is this the greatest distance at which these sounds might be heard?

The range of finback calls today is limited by background noise of similar frequencies—mainly the roar of ships' propellers. But finbacks have been evolving from other baleen whales for some 15 million years; their powerful calls must have been adapted to some purpose long before propeller-driven ships arrived.

Douglas Webb of the Woods Hole Oceanographic Institution and I have attempted to make a rough estimate of the range of the finback's call in the quieter oceans that no longer exist. By one set of calculations the sound under favorable conditions could carry 500 miles. This means that a finback could have heard a companion from anywhere within 785,000 square miles of ocean.

Any of our assumptions may be wrong. On the other hand, 500 miles may be far too low a figure. For instance, we assumed that sound energy from the finback spread out in the shape of a swelling sphere—a realistic assumption for short ranges. But different principles govern sound transmission in deep ocean; they may have interesting consequences for the lives of whales.

In deep ocean the speed of sound increases with temperature and with pressure; thus sound travels fastest near the surface and at the bottom. If we were broadcasting from the slowest, middle layer, sound rays directed almost straight up and down would reflect from surface and bottom and lose much energy. But oblique

The ocean gives a whale's voice enormous potential range. Diagram below shows how the finback whale's loud, low-pitched call travels in deep ocean. Nearly vertical sound rays, hitting sea bed and surface, soon dissipate but may reach nearby finbacks. Other rays avoid energy loss from reflection, traveling up and down over vast distances. They near the surface every 35 miles. Whales at intermediate points may hear a signal that bounces off sloping terrain—or they may miss many sounds. But if a call served only to signal the sender's presence, occasional contact would suffice.

Along such sinuous paths the finback's voice might have had a range of thousands of miles when no ships' noises interfered. More likely, says Dr. Payne, the finback would regularly have achieved 700 miles—sweeping an area equal to more than half that of the contiguous 48 states (above). Even in today's noisy seas finbacks have been heard at distances of 100 miles.

DIAGRAMS BY RICHARD SCHLECHT

rays under these conditions are always bent back toward the layer of slowest speed and do not lose energy from striking the ocean floor or ceiling. These bending rays would trace a sinuous pattern that weaves up and down across the middle. The sound would spread in the shape of an expanding cylinder, not as an expanding sphere. Channeled this way, sound travels much farther.

According to our calculations, if a finback were at optimum depth, prior to the intrusion of ships' noises, the maximum range of its call would be between 4,000 and 13,000 miles. A circle with a 4,000-mile radius covers about 50,000,000 square miles—some 17,000,000 more than the area of the Atlantic Ocean!

Though finbacks are not thought to dwell at depths required for maximum potential range, even a signal sent very near the surface can gain some benefit from cylindrical spreading. There is, however, no satisfactory way to estimate the farthest range of such a signal. We can only calculate the optimum case and then assume that the performance of finbacks lies somewhere between this value (4,000 to 13,000 miles) and the lower 500-mile maximum.

If the sounds of the finbacks do constitute a long-range signaling system, what possible advantage did these whales gain by developing a call that could carry across an ocean? The answer may lie in the finback's annual cycle. We know most about the Southern Hemisphere finbacks and will focus on them.

Although many people assume that all finbacks swim north to breed, there is good evidence that many never leave the Antarctic. Presumably some of the finbacks—juveniles, elders, new mothers —do not participate in breeding every year and need not make the full migration. At any rate the finback does not fit the familiar picture of a migratory species moving regularly between fixed locations. Everything seems to point to a sort of "emancipated migration" where destinations and participants may change dramatically. Plankton pastures, for example, may be frozen over one year, ice-free the next. Despite such fluctuations in migratory destination the herds concentrate each summer wherever food is abundant. For this, some sort of long-range communication seems necessary.

The finback's strange mating patterns also seem to demand such a system. Unlike animals that rendezvous during the breeding season at a fixed place, fin whales reach a breeding peak at the time of year when they are *least* concentrated. How can scattered animals pair up without some means of signaling to each other?

Generally we do not think of scattered individuals as members of a herd. Yet a lone whale may be in acoustic contact with others, may be part of a group that stretches (Continued on page 158)

In silhouetted grace dusky dolphins rally round a visitor to their emerald realm off Patagonia. Guided by converging seabirds, divers dropped in to find more than a hundred feeding dolphins. Sleeker and swifter than their ponderous right whale relatives, the dolphins cooperated to corral a school of fish. Observers could hear clicking noises like those reported from dolphins in echolocation studies. During such tests the animals could detect fish and avoid obstacles even when blindfolded.

Many people confuse dolphins with game fishes of the same name and also with the tubbier, blunt-nosed porpoises of the whale tribe. Some biologists favor the terminology of sailors; they call any small whale species a porpoise.

A specimen of the dusky was collected in these Patagonian waters during the epic voyage of the Beagle. Darwin called the species Fitzroy's dolphin in honor of the ship's captain.
WILLIAM R. CURTSINGER

A gallery of whales

Whales began to evolve from earthbound ancestors into sea mammals at least 60 million years ago. Today the order of whales, or Cetacea, numbers some 80 species; its members are warm-blooded, breathe only at the surface, and suckle their young. They range from 4-foot porpoises to the blue whale (1), as long as 90 feet, weighing 135 tons—the largest creature that has ever lived. Its heart circulates 2,000 gallons of blood; a child could crawl in its aorta. A newborn blue whale weighs two tons and may gain up to 500 pounds a day in its first week.

To swim, whales swing their flukes up and down—in contrast to the side-to-side motion of fishes. Slow in cruising, big whales may spurt briefly at 20 miles an hour. In speed tests a porpoise once reached 24.6 mph. The deepest diver, the sperm whale (2), plunges perhaps 4,000 feet, can stay under an hour or more.

The cetaceans divide into toothed whales that seize fish or squid, and baleen whales that strain food. Toothed whales have one blowhole, the others two. Furrows on the undersides of some baleen whales let the throat expand as water is scooped. In winter and in migration some species may live off their blubber for months.

The baleen group includes the giants. The cosmopolitan blue, fin (3), and sei (4) whales migrate to low latitudes in autumn, to higher ones in spring—the sei avoiding icy waters. The bowhead (5), or Greenland right whale, lives in Arctic seas; its enormous baleen grows up to 15 feet long. Three forms of the right whale (6), sometimes classified as distinct species, frequent cold Atlantic and Pacific waters. The scarce pygmy right whale (7) roams southern seas. The gray whale (8) is seen off California each winter as it migrates from the Arctic to breeding grounds in Mexican lagoons.

The little piked whale (9), flashing its white flipper patch, dwells in moderate and polar latitudes of both hemispheres. The humpback (10) grows the longest flippers—up to 14 feet.

Largest of the toothed whales, celebrated in *Moby Dick,* the sperm whale lives in all oceans. Bachelors may wander to frigid waters; harem herds remain in warmer climes. The great head yields spermaceti, long used in candles and ointments. The pygmy sperm whale (11), also widespread, has a more porpoiselike size.

The goose-beaked, or Cuvier's, whale (12), at home in temperate regions, and the bottle-nosed whale (13), with distinct northern and southern forms, have but a single pair of teeth. The male narwhal (14) of the Arctic usually displays only one tooth, an eight-foot tusk, its purpose a mystery; men once related it to the fabled unicorn. The beluga (15), or white whale, another northerner, resembles a tuskless narwhal. The pilot whale (16) is often seen off Scotland. Wide-ranging in deep water, the false killer (17) is akin, though not in repute, to the killer whale (18). Unjustly feared by man, the killer preys on fish, seals, penguins, and other whales.

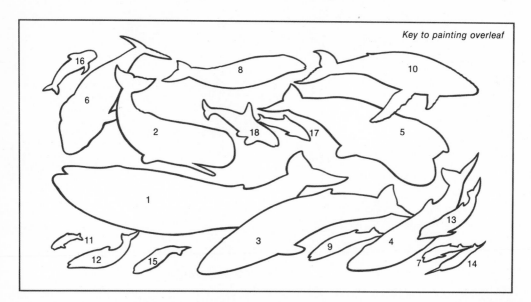

Key to painting overleaf

Flying flukes of a right whale slap the sea in a cetacean pastime called lobtailing.   154

WILLIAM R. CURTSINGER

PAINTING BY DAVIS MELTZER

far beyond the horizon. If so, a functional social group may once have covered a whole ocean basin. Perhaps we should start thinking in terms of a "range herd" in which a species could live in tenuous contact throughout large portions of its whole range, spreading out to find and exploit resources as they bloom and wane in the seas. Yet an animal that announces its presence across thousands of miles of ocean might attract predators. This way of life could evolve only in an animal relatively free of predators, a very large animal—a finback whale, for example.

This line of reasoning makes us want to know much more about the social structure of whales. But these creatures roam all the world's oceans, as much an embodiment of wildness as has ever existed. Even when you find them you may see little, for they are constantly vanishing beneath the shroud of the sea. We require some point of access where whales are abundant and resident before we can capture more than fleeting glimpses of their lives.

We have found such a place in Argentina. Off Patagonia a band of southern right whales appears each fall and lingers through the winter. A rare spot and a rare species. Once, right whales abounded off all continents. Whalers prized them for their oil and baleen and, dubbing them "right," slaughtered them without restraint. The

species today is perhaps the rarest of all wide-ranging mammals.

I shall never forget the sublime shock of my first visit to the Patagonian right whales. I found a 30-mile stretch of shore dotted every hundred yards or so with male elephant seals guarding their harems and adding females to them. Offshore, at times within a hundred yards, right whales patrolled at slightly longer intervals. Penguins scrambled in the surf; albatrosses, shearwaters, and petrels wheeled in the updrafts that form as sea breezes meet the land. Skeins of cormorants, gulls, terns, and sheathbills swept by. Far up the beach a colony of sea lions lolled.

Nearby rose sheer cliffs with layer upon layer of fossils — a myriad of extinct shellfish. Whale bones, too, indicating that whales have frequented these waters for hundreds of thousands of years. No wonder William Conway, director of the New York Zoological Society, who studied birds here, called the area "the greatest coming together of sea, land, birds, and mammals on earth."

In one bay, right whales swam so close to the cliffs that we could look down through the water and see them as if they were suspended in air. Around Bermuda I had lurched for days in a boat to

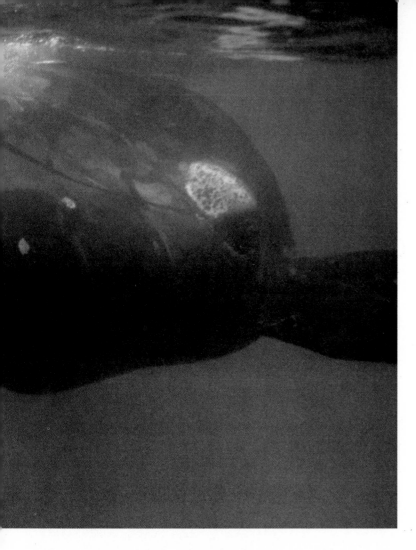

Collision course—or prelude
to mating? Though observers
could not pin down the sequence
here, the meeting resembled many
that preceded the onset of
courtship among right whales:
a male nudging close (like the
whale at right), an unwilling
female turning away. By rolling
on her back at the surface,
a female can completely frustrate
a male's advances. If she
decides to accept him, a change
of position leads to an embrace
with flippers, then mating.

Distinctive patterns of
callous-like growths on the
whales' heads enabled scientists
to identify individuals but not
their sex. With rare exceptions,
male and female could be
distinguished only by their
parental and breeding behavior.
WILLIAM R. CURTSINGER

catch an occasional look at humpbacks. Now I could sit in comfort and gaze at whales completely undisturbed by my presence.

The site was right and the whales were right, in more than just the name. Bulbous, callous-like growths—known to old-time whalers as "the bonnet"—adorn the heads of these whales. The callosity patterns vary with each individual. Many right whales also have distinctive white belly patches. Once you sort out the patterns, every time a right whale raises its massive head or rolls belly up it tells you who it is. Nothing quite like this has ever before been available for the study of whales in the wild.

In 1971, I returned for three months of research sponsored by the New York Zoological Society and the National Geographic Society. Day after day my colleagues and I observed the whales, swam with them, boated near them. Some shied away from us, others kept a discreet distance, a few curious ones let us come within arm's length. For the most part they were gentle. We recorded the head and belly markings of virtually every whale in the area.

We set up a triangular array of sonobuoys—floating transmitters with underwater microphones—to pinpoint the source of whale

sounds. A four-channel tape recorder picked up the sonobuoy transmissions and simultaneously recorded descriptions of behavior from observers on a cliff or in a light plane overhead. By studying associations between sounds and behavior, James Gould, a Rockefeller University graduate student, and I hope to discover some clues to the communication system of right whales.

Unquestionably the whales' chief social activity here is courtship and mating. The whales seem to be promiscuous. One may mate with two partners within an hour—the female breathing quietly at the surface, the male beneath, on his back, holding his breath. When a female wants to avoid a male, all she has to do to put herself out of reach is roll over onto her back.

At such times several males seem to try to push her down by overriding her. Eventually, of course, she must right herself to breathe. Then all the males dive amid much pushing as they try for a proper alignment. One would expect some violence in these melees, but we never saw any—sudden turns and hard bumps, yes, but no determined attacks. And whenever a female accepts a partner it is quite tenderly done, each hugging the other with its flippers.

Similarly a mother seems to exhibit tenderness toward her calf, holding it across her chest, or hugging and patting it. She will tolerate a good deal of shenanigans—bumping and butting, or the calf's swimming all over her back and tail while she tries to relax.

Once I watched a calf alternately breaching and banging into its mother. For half an hour she took it like a rock. Then on one pass she flipped onto her back. As the calf swam over, she caught its tail with a flipper and clamped it against her chest while the

Tête-à-tête at sea, man and whale take a close look at one another. Barnacles bedeck the right whale's "bonnet" and wraparound lower lip; a trickle spills down the lip, which covers baleen plates measuring as long as eight feet.

Undersea murk, blurring detail, turns a curious whale into a frightening apparition. Perhaps attracted by the noisy gurgle of breathing gear, the whale approached to within 15 feet. Its head arches upward, presumably to enable the eyes, set far back, to see ahead.

Only a few encounters proved hazardous. Flailing flukes accidentally cracked one man's ribs and flung him into the air. And once a whale made the sea boil with frenzied head-thrashing within a yard of a photographer. "It was the only threat of raw violence I have ever seen in this species," the author declared.
WILLIAM R. CURTSINGER

Right whales rocket into the air;
in moments they'll crash-land
with the boom of distant thunder.
Flipper reaching for the sky,
the leaper at right arches its back
as it turns on its side; the other
displays its white belly patch.
Both will land on back or side;
right whales seem to disdain
belly flopping.

Whales often breach; no one
knows just why. They may jump to
stun schooling fish, to shake off
parasites, to scan surrounding
waters, to show off to one another
—or just for the fun of it.

CHARLES NICKLIN, JR.

youngster struggled with choking breaths. Finally all calmed down; she slowly released her hold and they swam placidly off.

When a storm set adrift tangles of kelp, the whales had a new plaything. It was a delight to watch. A whale would lie on its side patting the kelp strands between its flippers, then come up underneath the kelp, letting it drape over the head like some fantastic hat. As the whale swam forward, the strands slid backward, and as they were about to fall off a swish of the tail brought them around where the flippers could catch and hold them, or just bat them about. Again and again the whales played this game, and always at the magnificent slow, slow pace that governs their lives.

Toward the end of our stay we noticed more and more mothers and calves concentrating in one place, while in another grown males were vigorously pursuing females, with few calves in evidence. From this we tentatively conclude that one area served principally as a nursery, the other as a mating ground.

From our hydrophones we heard a large variety of sounds occurring in many contexts—sounds like grunting, mooing, moaning, and sighing. Clearly there is a good deal of vocalizing among the mating groups. Occasionally a whale seemed to produce a kind of soft swallowing sound just before breaching. But firmer conclusions await a more detailed laboratory study of film and tape.

I plan to return for a much longer stay, in hopes of putting together a more complete picture of the life of this fascinating species. There is a bitter urgency about such plans; the whales are in danger. True, the voices of concern are rising, and factory ship whaling, with its deadly efficiency, is declining. But the killing goes on, and the industry pushes into new hunting grounds. Two companies operate in the South Atlantic—near our right whales—without heed to international restrictions. Both on occasion have killed whales of rare species for the production of dog food, automatic transmission fluid, cosmetics, and soap. Yet every ingredient taken from whales can be obtained from other sources.

It's all simply a matter of time, and then whales will be so rare that we will no longer be able to find them, even we who have received the grace of seeing them in their wild, free state. Maybe they will be unable to find each other. And the voices of whales will be lost in the relentless roar of ships' traffic.

Vapory geyser jets from blowholes of an exhaling right whale—a welcome sign of life from a species hunted to the edge of extinction. A century ago Melville wondered "whether Leviathan can long endure so wide a chase . . . whether he must not at last be exterminated . . . and the last whale, like the last man, smoke his last pipe, and then himself evaporate in the final puff."

WILLIAM R. CURTSINGER

THE STRATEGY OF THE NICHE

Gordon H. Orians

And NUH is the letter I use to spell Nutches
Who live in small caves, known as Nitches. . . .
These Nutches have troubles, the biggest of which is
The fact there are many more Nutches than Nitches. . . .
So each Nutch in a Nitch has to watch that small Nitch
Or Nutches who haven't got Nitches will snitch.

Dr. Seuss's little tongue twister has fascinated me ever since I first read it to my children years ago. The author is not an ecologist; yet his poem surprisingly captures the spirit of our current understanding of how organisms interact in nature.

We can define a niche as the relationship of an organism to a specific physical environment and to other species within that environment. But this merely describes the activities of a plant or animal—how it uses the resources of its habitat, how it may be used by other creatures. Ecologists want to know these things, but also much more. Evolution is a long-term process of niche-snitching, or switching, and what we want to know is, who is a good snitcher and why. Like the Nutches, ecologists have troubles, the biggest of which is, figuring out what a good—or bad—snitch is.

Niche strategy should interest everyone. For in this evolutionary game man is vitally involved, the strongest and most influential player, yet with no clear idea of the rules nor what effect his moves might have on the niches of living things—including his own.

It has taken a lot of travel and a lot of thinking to achieve even the limited understanding we have today. In East Africa, European explorers, and the tourists and trophy hunters who followed, marveled at the herds of big game. Ecologists, who comprise a particular class of tourist, wondered how semi-arid savannas could support so many large mammals of different species. Recent field studies have shown that some species concentrate on different grasses, while others share grass by cropping it in succession.

Other ecologists have discovered the fascinating diversity of communities at their very doorsteps. More than 200 kinds of butterfly and moth larvae attack the brown oak, the dominant deciduous tree in much of western Europe and similar to our white oak. Paul Feeny, now of Cornell, found that the caterpillars' numbers and life histories are strongly influenced by defensive chemicals produced by the tree, which inhibit the insects from using proteins in the leaves. In spring, when leaves are tender and chemical defenses low, the number of insect species is greatest; many of the larvae grow rapidly, completing their life histories on the leaves within a few weeks. When the leaves toughen and have more chemicals, the remaining species develop more slowly. Some take more than one summer to finish growing; others live inside the leaves, dining on the softer cells.

Clannish hyena, wandering wildebeest, territorial gull—for every species, survival demands a niche.   169

THOMAS NEBBIA

Savannas of the Serengeti nourish animals in breathtaking numbers and variety. In the green flush of Tanzania's rainy season, grazers like these zebras and wildebeests feast on tender short grasses rich in proteins, still low in tough cellulose. The dry months bring scarcity. Then each species relies on its unique way of life to get food adequate to its needs and nature. The three most numerous species — wildebeest, zebra, and Thomson's gazelle — divide the full-grown grasses by layers. Eating more and cycling food rapidly, the zebra fares well on the coarser top level, the flowering culms high in cellulose. The zebra's grazing and trampling help the smaller wildebeest, which takes the leafy middle layer, better for proteins. The little gazelle, most in need of quality rather than quantity, feeds closest to the ground — on the choice, protein-rich young shoots and on fruits of herbs.

A species' numbers do not fully gauge its impact on the plain. Ecologists also measure biomass — the total weight of a species in a given area. As the figures below indicate, elephants can have a greater impact than 17 times as many topi. The table, covering the Serengeti-Mara region of East Africa, shows the average population of the major plant-eaters and their total biomass in millions of pounds:

SPECIES	POP.	BIO-MASS
Wildebeest	239,500	98.9
Zebra	171,900	88.3
Thomson's gazelle	640,000	26.9
Topi	19,900	5.2
Kongoni	2,100	0.6
Eland	8,000	7.1
Domestic cattle	85,300	64.0
Impala	7,500	0.8
Waterbuck	1,200	0.6
Giraffe	2,700	4.7
Buffalo	21,800	28.9
Rhinoceros	200	0.4
Elephant	1,150	5.4
Totals	1,204,250	331.8

Among the first discoveries biologists made as they began to explore the world in the 19th century were the consistent patterns in the structure of natural communities of plants and animals. A grassland area anywhere in the world supports fewer than ten species of breeding birds; a forest nearby always has several times that number. Islands have fewer species than a comparable piece of the adjacent mainland, where no sea barriers limit the movement of living things. Plant communities at high latitudes, where wind pollinates the canopy trees, are consistently dominated by a few kinds of trees. In contrast, a wet, tropical forest, where wind does little to spread pollen and where trees must compete for animal pollinators, may have as many as 200 tree species to an acre.

The spectacular variety of tropical plant life extends to other organisms. A typical deciduous forest in North America provides suitable breeding conditions for about 35 bird species, half the number that nest in a similar forest of Costa Rica. There I studied forests that harbor, in addition to birds like those of our eastern woodlands, a bewildering array of fruit-eating birds, specialists on arboreal snails, and birds that sit and wait for some large nocturnal insect to reveal itself by an incautious move. It is the parrots, puffbirds, motmots, trogons, jacamars, and antbirds that lure us to the tropics, not the near-relatives of our robins, sparrows, and warblers.

Scientists have also wondered at the erratic fate of animals introduced to areas where they had never been before. Some quickly found niches free for the snitching, exploded in numbers, and became serious pests. Witness the starling and house, or English, sparrow in North America, the mongoose in the West Indies, the European rabbit in Australia. Other species failed completely, even after repeated introductions.

How to explain the consistent patterns in community structure around the earth, and the strange behavior of species in regions where they did not evolve? Ecologists felt the answer must be based in some degree on the assumption that organisms living together influence one another; that is, they interact in ways which cause some species to be more common in some places, to be entirely missing in others.

A nimals and plants can interact in only a few important ways. An organism can eat another, can cooperate with it, or compete with it. Predation is widespread, since animal life is a prime source of the preformed, energy-rich molecules which all organisms except green plants require as food. Whenever a potential food source is available, sooner or later some organism will evolve to make use of it. Who knows, perhaps the next spurt in evolution will bring forth plastic-eating bacteria which would find a major, presently unexploited food supply and become very abundant. Such welcome "predators" are unlikely to evolve in time to solve our plastic-disposal problem, but if we had the luxury of evolutionary patience a solution would certainly come.

Predation can strongly affect the niches in a natural community. In the rocky zone between high and low tide on our Pacific Coast the ocher starfish has such a major influence on the array of prey animals that we call it a keystone species. It prefers mussels but also eats barnacles, snails, and chitons. The mussels, the best competitors for space, could easily push the other shellfish out and dominate the entire zone — were it not for the starfish. They prevent the niche-snitching mussels from excluding the others, thus maintaining the intertidal community's diversity. Predation

172

(Continued on page 180)

JOAN AND ALAN ROOT

TO EACH A NICHE

Turtle and snakebird patrol their separate domains in Mzima Springs, a desert-rimmed chain of pools in Kenya's Tsavo National Park. Each must enter the other's element, the turtle for air and a place to lay eggs, the bird for fish, which it impales on its rapierlike beak. Thus do niches interlock in Mzima's miniature world.

In another pool, one fish snatches snails from the floor while another nibbles algae from the hide of a hippopotamus. A third gleans a living from the great beast's dung. In turn the fish nourish the snakebird, which gulps them whole, and the turtle, which paddles in for leftovers as a crocodile feeds. One day, bird or turtle may make a crocodile's meal—and nourish the fish, which will gather up scraps.

Lumbering lord of a crystal realm

Barrel-bodied leviathan of Mzima's transparent world, a hippo weighing two ungainly tons treads the bottom as if nearly weightless in a graceful slow-motion ballet. A retinue of *Labeo* fish cleans a hide whose white scars tell of this bull's duels with other males or battles with cows protecting their calves.

Hippos dominate the springs' ecology. With lips nearly two feet wide a full-grown hippo crops some 150 pounds of land grasses each night, then returns at morning and adds to the pools' rich carpet of dung—a nursery for bacteria, insect larvae, and crustaceans, from which countless food chains extend. This one-way flow of energy from land to water sustains a wealth of aquatic life unmatched in springs that lack a hippo herd.

JOAN AND ALAN ROOT

As a hippo moves, it stirs up food for fish and even for birds on its back. In death it leaves a ponderous bequest; slipping to the bottom a final time, it brings a feast to large and small.

Early in life hippos occasionally fall prey to crocodiles and lions, or a young one may be gored by the tusks—the knife-sharp lower canines—of its own rampaging father. Dangers diminish as its bulk expands; predators rarely attack an adult. Bull hippos sometimes fatally gore each other as they battle for the right to mate. Now and then a behemoth succumbs to one of the tiniest residents in its domain, a disease germ. Age fells many at perhaps half a century. Yet the herd lives on, numbering about 60, and bringing each morning its grist for Mzima's unceasing, superbly balanced biological mill.

The weak fall prey, the herd stays strong

In a rare drama a crocodile in Uganda's Murchison Falls National Park culls from a hippo herd an injured and suffering calf. Though hippo cows defend healthy calves, this one's mother turned away as the croc splashed in (opposite) and chased the calf to a sandbar.

Little changed from ancestors that preyed on dinosaurs, a growing crocodile expands its diet from insects and small invertebrates to fish, mammals as large as buffalo, and even its own young.

Hunting near the water's surface, a crocodile often floats log-like with only eyes, ear slits, and nostrils exposed. With jaws that can clamp down with perhaps half a ton of force, yet cannot open against a man's handhold, the croc seizes a drinking beast by leg or muzzle and drags it under to drown. Crocodile teeth hold like talons but cannot cut or chew, so the predator thrashes and rolls to tear off chunks it can gulp. Sated, it suns itself ashore, sometimes with its mouth agape. Into this awesome maw hops an occasional plover or sandpiper to pick the scaly reptile's teeth.

THOMAS NEBBIA

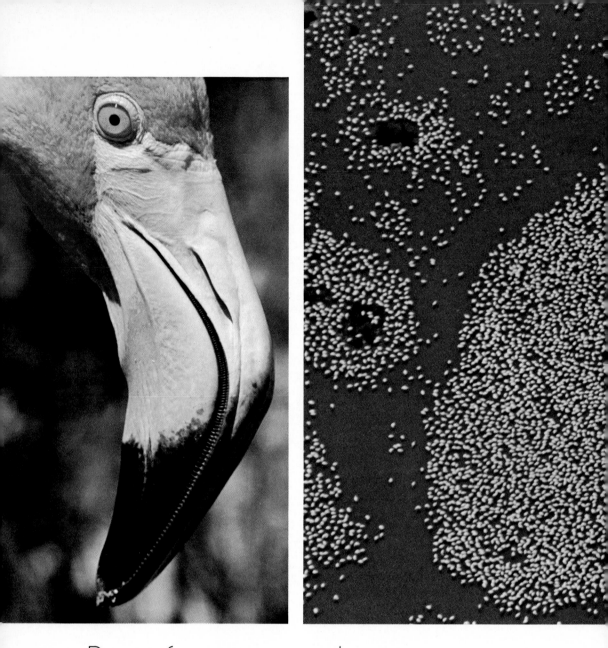

Room for many or only one

Lesser flamingos seem to fuse into living islands at Kenya's Lake Nakuru. Greater flamingos may also crowd in. The two species—one five feet tall, the other barely three—depend on different foods; the greater eats tiny aquatic animals while the lesser extracts its vegetarian fare of algae. An American flamingo (above left) shows the strainer bill that equips flamingos for their small-game hunt.

Bent on bigger game, a polar bear hunts alone off Norwegian islands. Unlike flamingos, which exist in a simple food web, this Arctic nomad tops a complex food pyramid that begins with plankton and includes fish and seals—the bear's favored prey. At each level only a fraction of the energy in the food source passes on; numbers decline as the pyramid builds, until it peaks with the lonely bear that must hunt over many square miles to meet its needs.

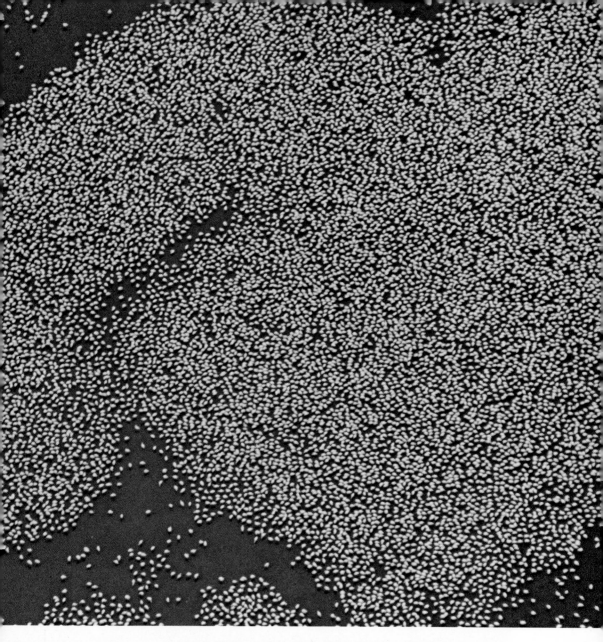

THOR LARSEN. ABOVE: GEORG GERSTER, RAPHO GUILLUMETTE. OPPOSITE: PAUL A. ZAHL, NATIONAL GEOGRAPHIC STAFF

can also keep species out of certain environments. Mountain sheep have been forced to live around cliffs beyond the reach of wolves. Many species of seabirds can nest only on islands where there are no mammalian predators to disturb the breeding colonies.

Some organisms can survive only with cooperation from others. Without protozoa or bacteria in its gut to convert cellulose to sugar, a termite living on wood would starve. Similarly, intestinal bacteria break down cellulose eaten by vegetarians like cows and elephants. Many plants devote energy to producing material useless to themselves but attractive to insect pollinators. And Nile crocodiles actually let spur-winged plovers — potential prey — into their mouths to feed on parasites among the crocodile teeth.

Ecologists have progressed most rapidly in understanding the nature of the third kind of interaction — competition. The aim has been to learn how closely related species influence

European rabbits encircle a water hole in Australia, where two dozen imports found no serious predators

each other when forced to compete for a vital resource. The first important study was made by the Russian ecologist G. F. Gause, who published his results in 1934 in an intriguing little book called *The Struggle for Existence*. He found that when he placed two species of single-celled paramecia in a culture with a single source of bacterial food one species always eliminated the other. Researchers could "fix the fight," choosing the winner by varying the temperature and other conditions of the environment.

From such experiments emerged Gause's Law, or the competitive exclusion principle. It holds that no two species requiring the same resources can live together if the resources are insufficient for both. Or, that no two species can occupy the same niche.

But the laboratory findings needed testing in the field, amid the more complex, widely fluctuating systems in nature. If Gause's principle were true, then similar species which

ompetitors and exploded into half a billion. Man stemmed the tide with fences, slaughter, and a lethal virus.

seemed to be ecologically identical, or sharing the same niche, should in fact have ecological differences which would prevent them from excluding one another. And investigators did find such differences.

David Lack of Oxford found major variations in diet among European birds of prey that lived in the same general area. Robert MacArthur studied the community of closely related warblers in our Eastern spruce forests. He discovered that each species prefers a different place in a tree and has its own manner of taking insect prey. The bay-breasted warbler works slowly outward from the shady interior, while the myrtle warbler flits from tree to tree, mainly near the ground. The Blackburnian and the Cape May warblers feed around the treetops, the latter moving vertically up the outer edges, the former outward along the limbs. The fifth member of the group, the black-throated green warbler, feeds at middle elevations, amid dense branches and around new buds.

In time many, but not all, ecologists came to believe in the importance of competitive interactions in determining the structure of natural communities, even though much of the evidence was indirect. For in most cases it is easier to measure the evolutionary results of competition than to observe competition directly.

As the processes of niche-snitching become clearer, we hope in time to develop a predictive theory of the niche that will tell us why species have come to do what they do in a total community. The door to such a concept was opened by that great pioneer of American ecology, G. E. Hutchinson of Yale. He suggested that we should determine what aspects of the environment an organism required, then measure how much of each resource the organism used. We would then have a description of the organism's niche in "n dimensions"—dimension being another word for the resource or aspect measured, and n representing the number of resources considered. Few of us, aside from mathematicians, can visualize Hutchinson's concept of the niche as an "n-dimensional hypervolume," but his work directed attention to two major ideas. It emphasized first the need to find important dimensions, and second the need to know how an organism spreads its activities along those dimensions. If we are interested in niche-snitching, we can ignore resources that are present in superabundance. Terrestrial animals, for example, never compete for oxygen, though all require it; for fishes, however, oxygen could be a relevant dimension, since oxygen supplies in water are limited.

Among animals, when one individual uses what another requires, the losing competitor doesn't merely sit around and watch. Often it fights for the things it needs—food, resting areas, good places to hide from predators. If two species have been interacting for thousands of years, their moves tend to become less and less direct. Rather than fight for a resource, they interact with respect to the space where the resource is found. We think this behavior evolved because it is often easier to defend a piece of ground than defend the resource directly; also because animals must often select living areas and begin reproductive activities before the resources are fully available.

This means that animals must be able somehow to tell from the appearance of an area something about its future abilities to provide the necessities of life. Most animals come to know their home range very thoroughly—where the food is, good escape routes to evade predators, good nesting sites. If an animal defends its home range from others of

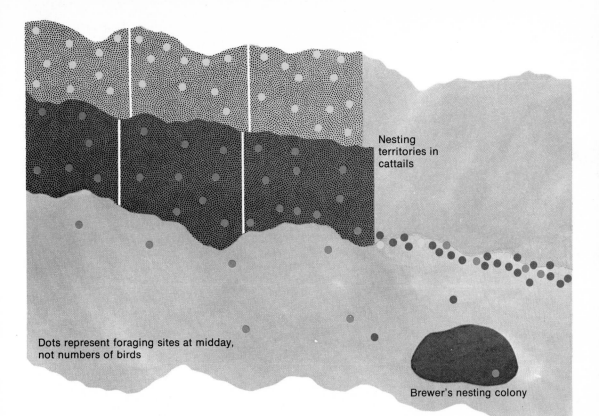

Nesting
territories in
cattails

Dots represent foraging sites at midday,
not numbers of birds

Brewer's nesting colony

● Red-winged blackbirds ● Yellow-headed blackbirds ● Brewer's blackbirds

*Ecologist in his habitat, Gordon Orians tracks
blackbirds in the Potholes of central Washington,
dry lava plains overgrown with grass and sage
and dotted with irrigation ponds. Vast swarms
of aquatic insects rise from the reedy ponds—
but not enough to feed all the nesting
red-winged, yellow-headed, and Brewer's
blackbirds. Parents forage all day, yet many a
nestling starves. How do these three similar
species interact to divide a limited resource?*

*Dr. Orians's team measured food abundance
by trapping insects as they rose from the water
and also after they flew upland. Collars kept
food in nestlings' gullets to gauge how much
parents delivered. Through the day, prey varied in
numbers and location; hunting tactics varied,
too, reducing competition among the birds.*

*Most fed by the pond when insect emergence
was heaviest (above). Yellowheads and redwings
hunted partly in nesting territories that held a
male and up to five females in each. The Brewer's,
poor hunters in tall reeds, deployed from their
colonial nest site to catch damselflies at the
grassy pond edge. Later, when prey on the
pond thinned out, yellowheads still fed there.
The Brewer's still preferred the edge but also
tried open uplands. And the versatile redwings,
good hunters on grass, bush, tree, or pond,
concentrated on the upland sagebrush.*

STEVEN C. WILSON. DIAGRAM BY TASI GELBERG PESANELLI, INC.

its own or different species, we call it territorial. The formalized defense of resources is one of the most conspicuous activities in many animal species. Though we may not recognize it as such, territorial behavior is often a form of niche-snitching.

Exploring the niches of a particular group of creatures, it's often said, requires the investigator to "think like the animals he's studying." For many years my friends have claimed that I was preadapted in this way for the study of birds. If so, I am indeed fortunate. Birds are conspicuous and familiar, and make good subjects for field research. More work on niche theory has been done with them than with any other group.

Fittingly enough, one of the most important early studies on the evolution of bird communities was done by David Lack on the finches of the Galapagos Islands off Ecuador—the strange creatures that had so intrigued Charles Darwin and had figured so prominently in his ideas on natural selection. The 13 kinds of finches there apparently evolved from an initial colonizing group of a single species. Some of them exploit the environment more in the fashion of warblers and woodpeckers than of finches.

Amid eerie creatures of the volcanic Galapagos Islands Charles Darwin first pondered problems of evolution. Here he met the world's only marine iguana; to him the seaweed-eating lizards were "imps of darkness." They are kin of the land iguana (opposite) on cratered Fernandina Island.

As expected, Lack found that bill size and shape were correlated with diet; species with stout bills ate large seeds, those with longer, thinner bills ate insects. More surprisingly, he concluded that a bird's characteristics were influenced by the presence or absence of similar species on the same island. Three ground finches showed the greatest difference in body and bill size on islands where they lived together. The variation was advantageous because they could divide the available seeds and berries according to size. Where the smallest species lived alone, it tended to grow bigger. It could freely choose larger food items, and for this broader diet a larger body and bill were better adaptations.

Science still studies Darwin's "most singular" finches. Three of these species—the large (above), medium, and small ground finches— have apparently influenced each other's sizes. They differ most where all forage together—each adapted to feed best on a size of seed less desirable to the others (below). But where the two

The study provided strong—though indirect—evidence of the effects of competition among the species. It also pointed to the fact that the division of available food is a very important dimension in determining the structure of the finch community.

smaller species occur singly and each can use all available food, both grow to about the same size—midway between the small and medium figures shown.

Robert MacArthur discovered another relevant dimension in his studies of North American bird communities in different types of vegetation. He found a strong correlation between the vertical distribution of the leaves and the diversity of bird life in them. A higher foliage spread in trees and understory offered more ways of foraging efficiently, thus held more species. In time MacArthur could predict the number of bird species if he knew only how the leaves were distributed vertically in a patch of forest.

Both of these men deeply influenced my own research. In the rugged Potholes region of central Washington State my students and I searched out the dimensions that account for the distribution of breeding yellow-headed, red-winged, and Brewer's blackbirds. All three kinds prey on the incredible numbers of aquatic insects, mostly damselflies, that emerge from the lakes and streams in spring and summer to metamorphose into adults.

We found that around noon, when the insects emerge in peak abundance, all three blackbird species feed around the water. At other times their foraging styles tend to diverge. The yellowhead, an aquatic specialist, remains around water all day. The redwing, a good generalist, moves to the uplands, where the insects go to rest after emergence. The Brewer's, with longer legs, cannot move around cattails well but patrols open ground more efficiently than the other two; it hunts over grassy openings and around pond edges which are clear of emergent vegetation such as cattails and bulrushes.

Knowing the kinds of food the birds eat and where and how they get it, we need to know little else about these species to understand where they will be found. Highly dependent on aquatic insects, yellowheads forage best over water, less efficiently in trees. Thus they have evolved to be "turned off" by trees; they always nest in emergent vegetation around lakes that produce the most insects. Redwings and Brewer's, however, nest over water, or in trees and bushes—even on the ground. But Brewer's never breed around lakes that are completely encircled by tall reeds.

We can even predict how the species respond to the space in which they breed. Redwings and yellowheads forage over an area whose shape makes it easily defensible by a

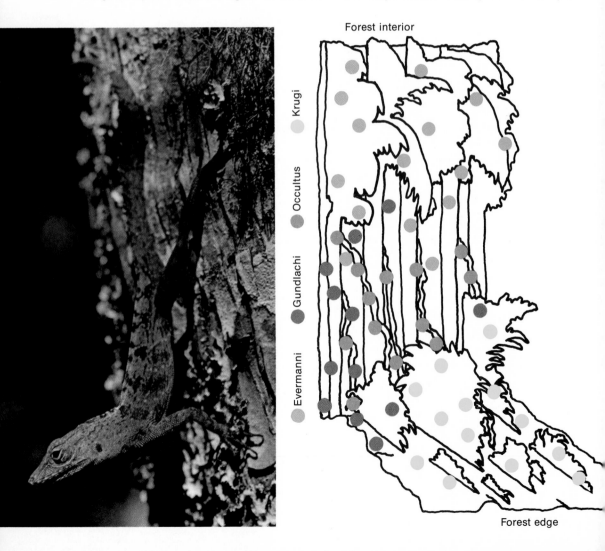

single male bird. Both species are strongly territorial. Brewer's, however, seeks food along the water's edge—a linear space much more difficult to defend. It seems better for a Brewer's to find the approximate center of the food supply and place its nest there. The same goes, of course, for all the Brewer's in the vicinity. This is probably what led to the habit of breeding in colonies.

Studies like these indicate that the "where" and "how" of food exploitation is an important determinant of bird distribution. And the n in the n-dimensional hypervolume for birds appears to be a very small number; we have to know only a limited number of things in order to predict the rest. This is exactly the result we must have if a general theory of the niche is to become a reality. But what of other organisms?

Like birds, lizards are mostly diurnal creatures, relying primarily on vision for locating prey, evading predators, and interacting with mates and rivals. Like birds, too, they are easy to study by a visually oriented animal such as an ecologist. Not surprisingly, we have learned a good deal about the niches of lizards.

In the West Indies the lizards of the genus *Anolis* (to which our American chameleon, or anole, also belongs) abound and have radiated into an amazing variety of forms. Unlike

Lizards find living room by clinging to distinctive perches in the wet, montane forests of Puerto Rico. All belong to the genus Anolis; all make a living by catching insects. But each of the four species frequents a different part of the vegetation—and each has a color scheme that helps hide it there. Green evermanni prefers high perches on tree trunks and in crowns. Brown gundlachi sits on trunks mostly below five feet. Krugi, striped yellow and brown, hunts on low leaves and stems. Grayish occultus, a rare form unknown to science until the 1960's, hugs twigs and vines; not much is known as to its height preference.

Sizes and habitats of Anolis species in the West Indies may vary according to the kinds and number that live together—an indication that the lizards have influenced each other's evolution. Zoologist Thomas W. Schoener, supported by the National Geographic Society, found that on some islands he could predict where in the vegetation a certain species could be found by knowing which others were present.

birds, these lizards do not take care of their young, so that all the energy committed to reproduction is in the form of food provided in the egg. Therefore the success of the females as insect hunters will determine how much energy can be mobilized for the eggs. Foraging sites that would minimize competition between the species should then be important aspects in the lives of these lizards.

The observations of A. S. Rand, now with the Smithsonian Institution, Ernest E. Williams, Thomas W. Schoener, T. Preston Webster, and other researchers in the West Indies indicate that groups of *Anolis* species have evolved to exploit different habitats, different structural positions within a single habitat, and different sizes of prey. Small lizards take smaller insects than their larger neighbors do.

Each species in a group of four found in the high, humid forests of Puerto Rico tends to favor a different kind of perch; one prefers grass and bushes, another vines, a third the tree trunk near the ground, a fourth the upper trunks and crowns. The highest percher, *Anolis evermanni,* is often found with *Anolis stratulus,* the tree-crown species of the lowland forests—but *stratulus* is a much smaller lizard. In no case do two species of similar size and similar preference in perches occur together regularly.

In many ways our efforts to understand the niches of different organisms appear about as unrelated to human problems as they could be. The future of man does not seem to depend upon our knowledge of blackbird foraging or the diameters of lizard perches. Nevertheless, many of us believe that this kind of information may prove vitally important to us in perceiving the consequences of our intervention in the structure of natural communities. As yet we have no way of predicting the long-term effects of removing diverse natural vegetation from much of the earth's surface and replacing it with single-species stands of crops. And we still don't know the outcome of many of our ambitious experiments in transporting species into communities where they have not previously existed. In short, we don't know our own niche and how much we dare tamper with it, and yet we continue to tamper with it at an ever-accelerating pace.

What will be the results of such assaults on the integrity of living communities as those in South Viet Nam, where we dropped more explosives than fell on Europe in World War II and where nearly two million acres of forest were destroyed? It is perhaps ironic that tigers benefited from the war; they had about a quarter of a century to learn to associate the sounds of war with the availability of dead and dying animals, including human beings. It appears that the tiger population has increased in Viet Nam, in much the same way that wolves increased in Poland during World War II.

Clearing land, introducing new species, fighting wars—all man's activities that disrupt ecosystems are done with the expectation that they will improve life for at least some part of the human species. But as ecologists and behaviorists gather knowledge of the ways of living things, some of us have an uncomfortable feeling that *Homo sapiens* may be in danger of becoming the first species to snitch its own niche.

A strange truce lures a cleaner shrimp to the very jaws of a California moray eel—a prey animal freely approaching a predator in a cooperative interaction that benefits both. As the shrimp picks tiny parasitic crustaceans from the fish's skin, the cleaner gets a meal, the eel a cleaning—unless,
188 *in rare moments, the moray prefers a morsel of shrimp and reverts to its predatory mores.*

THE
ELEPHANT:
LIFE AT THE TOP

John F. Eisenberg

We picked up their trail where it crossed the road and scouted the length of it, a 2½-mile march between water holes. We counted the footprints of a herd of 12, adults in the van, younger elephants behind. They had passed through the day before; scarab beetles had already started to attack the dung piles. And they had moved directly, for I could find hardly any sign of feeding—one scuffed area, eight broken branches.

We sniffed elephant odor on a few "rubbing trees," and near the road we came upon a digging spot where the herd apparently had paused for a dust bath. At times the trail narrowed to 16 inches, but the trees to either side stood far enough apart to let the animals slide through easily. They had trampled one spot heavily; here, perhaps, a calf had rested while the others milled about.

They had milled and defecated at three other places, always at forks in the trail. There may be a touch of irony in that. These elephants represented the largest terrestrial mammals in existence—powerful, intelligent, protected within the boundaries of Wilpattu National Park in Sri Lanka (then Ceylon). Yet the simple matter of deciding which fork to take had produced telltale signs of stress.

Tracking an elephant herd can be an extremely instructive pastime. Among the shy, nocturnal forest herds of Wilpattu, tracking may often be the only way to fill in details of herd movement and feeding habits—basic clues in the study of the ecology and behavior of elephants in the wild.

As a mammalogist I have been committed to the search for an understanding of mammalian social systems. I began with rodents and insectivores (partly a practical consideration, since you can keep a relatively large number in a small space). Later, as a zoo scientist, I became engrossed with the interaction patterns of elephants and other large mammals—and realized how much we had to learn about them. The emphasis on primate field studies in the

Looming bulk of an African elephant breaks up a banquet on the savanna. Clenching her moveable feast, the lioness retreats to a more tranquil spot. The elephant's alert stance and spread ears enhance the image of frightening size; relaxed trunk indicates only mild concern. Before a charge the trunk tucks under in a curl. A vegetarian with no interest in the carnivore's kill of kudu, the elephant merely seeks to shoo her from a nearby water hole.

With their enormous appetites, elephants prune forestlands, crop grasslands, spread food and seed in quantities that can quicken the rhythm of nature. In strength the adults stand unchallenged; of all the animals they encounter, only man, with his weapons and encroachments, can threaten the elephants' mastery of their domain.

JEAN-ERICK PASQUIER, RAPHO GUILLUMETTE

1950's and 1960's only underscored the need. I was convinced that science couldn't correctly interpret primate societies until it had comparative research on a whole cross section of mammalian species exploiting the environment in different ways. When the opportunity arose to observe elephant behavior on Sri Lanka, I took it. Field knowledge of this important animal was lacking, even though it has been working for man for thousands of years.

The two living species of elephants—the African elephant of the lands south of the Sahara, and the Asian elephant of India, Sri Lanka, Burma, Thailand, Malaysia, and Indonesia—today represent an end point in a lineage that began some 40 million years ago with the pig-size *Moeritherium*. Through the ages probosocideans evolved in a variety of forms, usually ponderous and endowed with the versatile trunk for which this order of animals was named. Mammoths and mastodons were staples of early man's diet. Both of these forms migrated across Bering Strait to the New World long before aboriginal man did, only to disappear under the pressure of predation and climatic changes.

The history of the taming of elephants is much more recent. Some evidence suggests that men trained elephants in the Nile Valley in the time of the early pharaohs; in Asia domestication appears in the Indus Valley around 2000 B.C. Later, elephants won renown in the armies of the Greeks, Carthaginians, Romans, Persians, and numerous kingdoms of India and Southeast Asia.

Intimately interwoven in the religious lore of Hinduism and Buddhism, the elephant holds a proud place in Eastern festivals. In Sri Lanka's colorful Kandy Perahera the symbolic tooth of Buddha himself rides in splendor atop an elephant with gold-sheathed tusks. The beast of ceremony bears more mundane burdens as well, harvesting hardwood in dense forests of tropical Asia, where the relatively few trees of commercial value make road-building economically impractical. And around small sawmills the elephant trunk—that wondrous tool that can sniff danger and trumpet a warning, serve as shower head or snorkel, snap a twig or delicately enwrap a fallen berry—lifts and hauls massive logs.

The elephant has a number of behavioral tendencies which are helpful for taming. As a herd animal it has a great need for social contact. A knowledgeable handler can fulfill such needs; aware that elephants often place trunk tips within each other's mouths, he will "greet" his charge by placing a hand in its mouth. Life in a herd also adapts the elephant to what psychologists call negative conditioning. A pesky youngster knows the instructive value of a

Bewildered Asian elephants churn the jungle dust of India's Mysore State as they flee from the din of bamboo clappers and men's shouts. Neither brute strength nor keen intelligence nor the herd's marvelous social structure will avail—for these cows and their offspring the way leads relentlessly toward a khedda, *or stockade. Then begin months of training for animals already attuned to learning by their own life-style. In elephant society youngsters learn the lore of their clan from cows who guard and nurture them (opposite).*
HARRY MILLER

193

swift kick—or a smack across the buttocks delivered by the multi-purpose trunk; and a dominant elephant reinforces its status in a herd by pushing subordinates around. In the same way, a trainer maintains dominance with occasional mild punishments.

Then there's the fabled elephant memory. The old saw that "elephants never forget" is of course an exaggeration. But the elephant does have a remarkable ability, under certain conditions, to remember activities expected of it so that it can perform complicated tasks with a minimum of control. The size and structure of the elephant's brain suggest a considerable potential for information storage and retrieval. True, the ratio of brain to body weight is less than those of the gorilla and chimpanzee, but we really don't understand the significance of such ratios. What does stand out is that the elephant's brain—9½ pounds in the average adult Asian female—is more than three times as large as a man's and that the elephant's memory and ability to manipulate its environment are certainly equal to those of the higher primates.

But the elephant's nature has some drawbacks which have prevented total sustained domestication. Bulls tend to grow aggressive and unwieldy during their periodic rutting phase and must then be kept isolated and in chains. Hence most of the work force consists of females. Expectant mothers in the late stages of a 22-month pregnancy—and new mothers—must go on light duty. Elephant men have often found it cheaper to replenish domestic stock by capturing wild elephants than by breeding them in captivity.

In the wilds the elephant on its daily rounds touches the lives of many species. In Sri Lanka we spotted mongooses at water holes dug by elephants in sandy stream beds during the dry season; it's likely that deer, monkeys, and other mammals drank there too. On trails maintained principally by elephants we met buffaloes, boars, leopards, mongooses—and men. The elephant is an ecological dominant. Though it may be one of the least abundant animals in its range, in terms of its biomass (the total weight of elephants per unit of geographical area), it can be the greatest contributor to the cycling of plant life.

The elephant represents the ultimate in adaptation for feeding on coarse plant material. Foraging 16 hours a day, the adult Asian elephant consumes up to 400 pounds of leaves, fruits, shrubs, grass, and bark. A sloppy eater, it leaves forage for other browsers that can't match its 10- to 12-foot reach. In 24 hours ten elephants can deposit more than a ton of feces on the forest floor. Butterflies and beetles feed on the dung, jungle fowl retrieve seeds from it, mushroom and other fungus spores thrive in it. Termites, whose importance in the

Braced by tusk and pillarlike forelegs, a mahout in Sri Lanka brushes his elephant's teeth with a coconut husk. Bleached hide, common in elephants of Sri Lanka, results from friction or comes naturally with advancing age. Routine maintenance like this, plus a few hundred pounds of bark and leaves each day, pays off in a jungle bulldozer that can operate for half a century in trackless forest, river, and swamp impassable by machine.

The African elephant— taller, bulkier, and bigger-eared than the Asian species—bears an undeserved stigma as a difficult animal to tame. Modern trainers in the Congo have demonstrated its skill as a lumberjack, and history celebrates its feats in war and pageantry. Hunters still stalk it for meat, for trophies, and for ivory— treasured for at least 8,000 years as a raw material of art.

GILBERT M. GROSVENOR, NATIONAL GEOGRAPHIC STAFF

food webs of tropical ecosystems we often underestimate, break down the undigested cellulose which comprises much of the elephant's waste. Like other higher animals, the elephant has no digestive enzyme for converting cellulose to nutrient sugars. The caecum, an enlarged pouch in the digestive tract, serves as a "fermentation chamber" for bacteria and protozoa that can reduce some cellulose, but most of it passes through. Termites attack the cellulose in dung and convert it to sugars. And such anteaters as the pangolin and the sloth bear dine on termites. Thus the elephant's activities serve to increase the flow of nutrients within the overall ecosystem.

In the main, predators find slim pickings among elephants, whose great size enables them to hold their own with all but man. An encounter may provoke little more than casual notice. One afternoon in Wilpattu I saw a herd of six, led by an old cow, approaching the spot where a leopard was lying near the edge of the forest. As the lead cow got to within 20 feet, she tossed her trunk toward the leopard, expelled a blast of air, and continued on her way. The leopard rose and slowly walked into the woodland.

Of course a lone calf or juvenile is highly vulnerable, but the elephant social organization supplies both protection and the milieu in which the young animal can mature and learn its role in life. This is true for both species of elephants. The observations my colleagues and I made over a period of three years in Wilpattu and elsewhere in Sri Lanka show a remarkable similarity to those made in recent African field studies.

The cow herd in which the young grow up serves as a repository of traditional knowledge vital for survival: the routes to water holes and feeding grounds, the seasonal movements to new ranges. It is possible to build and transmit such a repository because the elephant's large brain enables it to learn by imitation—and to remember. Also, since the elephant is potentially a long-lived animal and matures slowly, several generations can exist in the same herd, increasing the young's opportunity for learning.

The birth of a calf stirs great excitement in a herd. Indeed, the presence of several infants can induce lactation among cows without nursing young of their own. This helps ensure adequate nutrition for the young, because a calf can nurse from more than one cow and need not depend solely on its mother for milk. For the first six months a mother is extremely vigilant as she tends her calf. In the next half-year the calf begins to experiment with feeding itself, imitating elders in food-gathering and selection, "stealing" food with impunity from its mother's trunk or mouth. It may continue to suckle well into the second year, weaning gradually as it feeds more and more independently.

For the first four years of their lives both male and female live within their herd. This social unit is essentially a matriarchy led and coordinated by the oldest reproductive cow, which is probably mother, aunt, or grandmother to all the other adult females present. On Sri Lanka such clans or cow herds number from 8 to 21 animals.

By day the herds split into smaller subgroups for feeding, reassembling at a watering place in the evening to drink—each adult siphoning up as much as 15 gallons. At night forest herds move into the open to graze until dawn. Grownups generally sleep standing; even when a herd does lie down, some cows remain upright in a vigilant stance.

Elephants love to bathe, showering themselves with water and afterward with dust. The coat of dust may reduce insect bites and probably provides a comforting shield against the

Feasting in tropical forests, elephants spread sustenance for a host of neighbors. In the lowlands of southeast India and Sri Lanka, termites (1) tunnel to twigs dropped by the feeding elephant. Bacteria and protozoa in termites turn the cellulose into sugars useful to the abundant earthbound insects, on which the sloth bear (2) and pangolin (3) prey.

Juices of elephant dung feed fungi (4) and butterflies (5). Bits of feces make nests for the eggs—and food for the larvae—of dung beetles (6). The beetles make a meal for the jungle fowl (7) and mongoose (8)—which eats the bird as well. Termites attack cellulose in dung; so do soil bacteria (9), reducing such organic matter to fundamental nutriment for the plants on which the elephant thrives.

heat of the sun. Elephants also use their fanlike ears for cooling. On Sri Lanka we kept a telescope on a feeding group, counting and timing the flapping of ears under varying weather conditions. When the heat intensified (through a decrease in cloud cover or wind velocity), so did the flapping—sending a cooling flow of air over blood vessels that lie close to the surface of the ears.

The herds of Sri Lanka frequently moved to a new water hole every day, completing a circuit of drinking sites and adjacent feeding areas every three weeks. Between dry and rainy seasons they made longer treks of perhaps ten miles to a new feeding range. The daily routes often seemed random, but the home range for a given season remained remarkably consistent. In some areas herds cropped

Majestic bull, roaming alone
as grown males often do, strides
a rain-laced clearing in Kenya.
The curved forehead identifies
the African male; cows show an
angular profile (page 204).
In Africa both sexes usually
grow tusks. In Asia a tusker
is almost certainly a male.

Ambling easily (like the bull
shown here) or running at speeds
up to 20 miles an hour, elephants
use but a single gait—a modified
pace in which the legs of each
side move in sequence: left hind,
left fore, right hind, right fore.
The burden of supporting such
great weight rules out trotting,
cantering, galloping, or jumping.

Wandering alone or in small
groups led by a senior sire,
bulls cross paths with foraging
cow herds, breeding when they
find cows in heat. Inveterate
loners—sick, sexually frustrated,
or old bulls pushed out of
dominant status—may turn
into rogues and marauders.

The elephant can live in widely
varying habitats. Some observers
believe few other mammals besides
the baboon and man can match its
adaptability. Before men hemmed
them in, elephants made seasonal
migrations of hundreds of miles.
Herds still rendezvous between wet
and dry seasons—some to begin
short treks to new range, others
merely gathering in a vestigial
pattern of migratory behavior.
BRUCE DALE, NATIONAL GEOGRAPHIC PHOTOGRAPHER

the same shrubs over and over, maintaining their height with all the regularity of a gardener tending a formal hedge.

As a cow herd divides during the day, "nursery units" of mothers with small calves tend to separate from females with half-grown young and from males in their sixth or seventh year. This has obvious advantages: The older, more mobile animals roam farther, reducing competition for food with the nursing mothers tied down by suckling infants. The same advantages accrue from another aspect of elephant social behavior: Grown bulls are semi-solitary, with their own movement patterns; thus they do not compete for the protein-rich grasses required by females and their growing calves.

While females remain with their herd for life, males begin to

At the season's turning, herds of elephants convene, as did their ancestors in the migrations of yesteryear.

PETER BEARD

Here they trample an African savanna, milling and trumpeting, apparently spooked by the aircraft overhead.

diverge around their fifth year. As they grow, they develop a dominance order with pushing contests. Usually one gives way quickly, with no damage done. On rare occasions the loser will get ripped by a tusk. Gradually the males drift away from the cow herds to follow and feed in the vicinity of older bulls.

Between the ages of 14 and 17 the Asian male first comes into musth; that is, he begins to secrete a smelly, sticky fluid from the temporal glands located on either side of his face between the eye

and ear. The condition, marked by enhanced aggressive and sexual behavior, may be likened to the rutting season of deer and antelope. The bull elephant need not be in musth to breed, but it probably helps him to achieve dominance among competing bulls as well as among prospective breeding partners.

A cow herd, with a much larger home range than any given male's, will ordinarily pass through feeding areas used by several bulls. Since cows come into estrus about every four weeks, ample

*Maneuvering for mutual aid,
a family group forms a classic
defense cluster on the scrubland
of Tsavo National Park in Kenya.
Grownups ring their young;
an old matriarch anchors the
perimeter, ready to threaten
or charge—or merely stand alert
while the photographer makes
his picture and drives off.*

opportunity exists for a bull to join a herd and travel with it while a cow is receptive. Several other males may join the herd at such times; then tests of strength adjust the dominance order. Among elephants might makes right.

As courting begins, bull and cow sniff each other and touch trunks to mouths, genitalia, temporal glands, feet, and tails. They may pause to break branches or plaster their bodies with mud. At times they stand head-to-head, trunks intertwined, touching mouth to mouth. Or the bull may reach over the cow's shoulder with his trunk and try to bite her neck. During the mounting sequence other cows and juveniles may form a ring around the mating pair.

While courting, elephants communicate with visual, olfactory, and tactile signals. Without communication, social life is impossible. The temporal fluid of a male in musth is an agent of olfactory communication; males rub it on tree trunks, leaving a sign of their presence and mood. The African elephant also marks trees with such secretions, but in that species both male and female emit temporal gland fluids and—in contrast to the Asian male—there may be no relationship between the male's glandular activity and his state of aggressiveness or sexual arousal.

Elephants also use a form of communication more familiar to man: sound. They have a large repertoire of growls, roars, squeals, trumpetings, and snorts—for warnings, greetings, distress, locating straying calves, signaling attacks, or just keeping in touch. When an aroused elephant bounces its trunk tip on the ground during a snort, you can sometimes hear the resulting boom a mile away.

Death watch: A sick cow slumps to the ground; in seconds the social ties of a lifetime bring the families of her herd screaming to her side (upper left). Now—before the astonished eyes of ecologist Harvey Croze, working in Tanzania's Serengeti National Park—begins a drama as poignant as any in the children's tales of Babar:

A bull lays tusks against her, struggles to lift her; the burden, perhaps four tons, is too much. He yanks up a trunkful of grass, places it by her mouth (opposite). Then—"We can scarcely believe what we see"—the bull mounts her, as if trying to breed. All in vain. With a lurch she heaves over, dead.

For hours the groups drift off and return to the body, now grown pale (above). The bull keeps trying to mount. At last only a huge cow remains. She trumpets; the distant herd answers. She nudges the lifeless elephant one last time, and plods slowly into the dusk— toward the society of the living.
HORST MUNZIG

Clouds of dust rise from the plains above a herd of feeding elephants. Forelegs kick back and forth, back and forth, scuffing up clumps of grass. Each clump goes into a pile; when this grows big enough, the elephant seizes it with the trunk, strokes the grass up and down a leg to scrape off the soil, and eats.

Such sights are common during a drought. At such times I have also seen elephants push over trees to gnaw on roots or to bring high leafy limbs within reach. They can have a tremendous impact on vegetation. In arid parts of East Africa, as fire and elephants turn sparse woodlands into semi-desert, grazers like the oryx and zebra manage to hold their own. But browsers like the lesser kudu and rhinoceros go hungry.

Elephants, of course, can live in balance with their ecosystem, contributing to the cycling of plant life, helping many other creatures make a living. But in optimal habitat elephants rapidly increase and soon begin to render their range unfit for their own occupancy. In recent years the idea has grown among scientists that long ago, before Europeans began to exploit the continent of Africa, elephants utilized their habitats in cyclical fashion. Over a period of several hundred years the populations would build to a peak, crash, and disperse into new areas.

Today man, the elephant's ancient competitor, has blocked such options. As elephants increase in number, they strain the capacity of the preserves allotted to them. Observers in Africa have found that among dense populations youngsters take longer to mature sexually; the birth rate drops. The destruction of habitat cuts down shade and exposes calves to the debilitating heat of direct sunlight —opening the way to higher infant mortality. Zoologist Sylvia Sikes has recently noted a high incidence of stress-related diseases in extremely dense, disturbed populations in Africa. Hungry herds, ignoring park boundaries, raid neighboring farmlands, destroying crops and terrorizing villagers. I have seen the damage elephants can do, and the damage done to them—animals gunned down, many others carrying visible wounds.

In Africa the elephant's plight has received widespread attention; at least a start has been made toward effective management of the animals and their environment. But in India and Southeast Asia the whole problem of wildlife conservation remains in a curious state of limbo. Unless scientists and government officials make a concerted attempt to establish some management baselines for the existing preserves, I can only conclude that the elephant—along with such other spectacular beasts as the gaur, rhinoceros, nilgai, and black buck—will pass beyond the point of no return before the end of this century.

Another tug, another snack of bark, another tree stripped in Lake Manyara National Park, Tanzania. The elephant begins by digging into the tree and levering up the bark with the servant tusk—the stronger of the two, favored much as humans depend more on the right or left hand. Then the trunk takes over.

The scene symbolizes the danger clouding the elephant's future, even in sanctuaries like Lake Manyara. Protected herds multiply, tipping the delicate ecological balance. Food supplies dwindle. Then preserve managers in Africa and Asia face a sad dilemma: fight an uphill battle for more land, watch animals sicken and waste—or kill off excess numbers to match elephant populations with available habitat. In Lake Manyara, where bark-stripping could wipe out woodlands, culling may destroy hundreds of elephants.

THOMAS NEBBIA

LIVING WITH
MOUNTAIN GORILLAS

Dian Fossey

ightly at first, two hands pressed down on my shoulders. They belonged to Bravado, a mountain gorilla living on the slopes of Mount Visoke in Rwanda. Minutes before, Bravado had left his group, feeding some 50 feet away, to climb into the tree in which I sat. Brushing past as if I were not there, he had climbed about 15 feet above me to have a look around. Now, coming down from his sentry post, he had decided to assert his importance. Hand pressure is a signal one gorilla often gives another to claim the right-of-way on a narrow tree trunk.

I chose not to yield. Tightening my footlock on the branch, I awaited the young blackback's next move.

The pressure on my shoulders—surely only a minimum expenditure of a male gorilla's strength—continued for a moment. When I did not respond, Bravado moved back up the trunk and stood erect beating his chest. He jumped out onto a limb and hung by his arms. Then, much as a frustrated human might slam a door, he began bouncing deliberately, knowing that the branch would break and create a loud, sharp, satisfying noise. The branch gave way and he landed with a thud—unhurt—eight feet below.

Alarmed by the noise, his group rushed toward us, following the dominant male I call Uncle Bert. The leader's loud roar and the screams of others created a frightful din. But when they found Bravado quite intact and feeding, the group immediately settled down around my tree to eat for another two hours, almost as unaffected by my presence as if I were one of them.

This familiarity was not easily won. For the past several years I have spent most of my days with mountain gorillas. Their home, and mine, has been the misty wooded slopes of the Virunga Mountains, eight lofty volcanoes—the highest 14,787 feet—shared by three African nations: Rwanda, Uganda, and the Republic of Zaire (the former Democratic Republic of the Congo).

Climate of trust enables Dian Fossey to live and work among central Africa's mountain gorillas, whose fierce appearance belies their shy and gentle nature. At first, upon spying Miss Fossey, the powerful primates would scream and run. But she learned to put them at ease by mimicking their gestures and vocalizations. Her folded arms, a submissive gesture, show she means no harm.

Miss Fossey's long-term study of gorilla habits, numbers, and ranges is designed to help save this largest of the great apes, whose habitat is threatened by the encroachment of man.
BOB CAMPBELL

209

Veiled in forest fog, a young
gorilla plucks edible vines that
entwine a Hypericum tree.
Gorillas spend most of their
time on the ground, but will
climb to play, nap, feed—or
get a better view. Full-grown
males seldom venture up a tree.

At home in the highlands
near her subjects, the author
occupies this two-room
sheet-metal cabin. One of four
Africans on her staff tends
the pit fire used to dry clothes,
heat water, and provide warmth
in the chilly air at 10,000 feet.
BOB CAMPBELL

Gorillas, among man's closest animal relatives, roam the slopes and saddles in groups. The mountain subspecies, akin to the lowland gorilla of western Africa, is the largest of the great apes. A full-grown male may be 5½ feet tall and weigh 400 pounds or more; his enormous arms can span 8 feet. Though the gorilla is not an aggressive animal, its impressive size and awesome displays enable it to hold its own against all enemies but one — man.

The mountain gorilla's range is limited to small tracts of lush wet forest in central Africa. There only a few thousand remain, leading a precarious existence. Part of the territory they occupy has been set aside as parkland. In theory, gorillas are strictly protected; in fact, they are being pushed into ever smaller, more remote ranges, chiefly by herdsmen, farmers, and poachers. Unless a more determined effort is made to save it, the mountain gorilla will cease to exist in its natural habitat in the next two or three decades.

One of the basic steps in saving a threatened animal is to learn more about it: its diet, life processes, range patterns, and social behavior. To this end, I began my long-term study in Africa, with support from the National Geographic Society.

My camp, situated at 10,000 feet on a 12,175-foot mountain, lies within the Parc des Volcans in Rwanda. This tiny agricultural nation, its land heavily cultivated and overgrazed, is the most densely populated in Africa. The gorillas in the park had been so harassed by herdsmen and duiker-hunting poachers that the animals rejected my initial efforts at contact.

I too was an intruder, and they let me know it. In my first few weeks with the gorillas I heard more high-intensity roars, *wraaghs,* and screams than the peaceful sounds they commonly make when their daily life proceeds undisturbed. For my studies to progress, the great apes would have to become habituated to me — comfortable enough in my presence to behave normally.

To elicit their confidence and curiosity, I tried acting like a gorilla. I imitated their feeding and grooming, and later, when I was more confident about what they meant, I copied their vocalizations, including some startling, deep belching noises. I found that I could slow down the flight of unhabituated animals and even induce them to return by imitating their activities. Admittedly these methods are not always dignified. One feels like a fool thumping one's chest rhythmically or sitting about pretending to munch on a stalk of wild celery as though it were the most delectable morsel in the world. But the gorillas have responded favorably.

Several groups now accept my presence almost as a member. I can approach within a few feet of them, and some of the animals, especially juveniles and young adults, have come close enough to touch my clothing or equipment. I know many of them as individuals, each with a distinctive personality. And, mainly for identification in my notes, I have given them names.

Early in the study I learned to use the gorillas' great curiosity about my actions. I had accidentally startled a group at very close range, causing them to retreat into the foliage. Not wanting to disturb them by following—but anxious to see what they did—I tried to climb an unclimbable tree. While concentrating on this slippery challenge, I suddenly had the eerie sensation of being watched. Slowly I turned to confront a row of gorillas obviously intrigued by my clumsy efforts. Some were even leaning forward in their intense curiosity. Hoping to hold their interest, I attacked the tree for another ten minutes and at the same time managed to make my first accurate count of this group.

The range of the gorillas within the Virunga Mountains spans an east-west distance of about 25 miles and a width of from 4 to 12 miles. Gorillas are not found on the two active volcanoes of the chain, where vegetation is drier and scrubbier, but they do live on the six dormant ones. Here in woodlands, where succulent vines and herbs are plentiful, dwell an estimated 400 gorillas. On Mount Visoke each group studied has a home range of approximately five square miles. They travel a part of it each day and bed down at dusk, but they do not wander at random. They repeatedly use select trails and favor certain nesting and feeding spots. In my study area, the gorilla ranges are not exclusive but overlap.

Duikers and buffaloes also roam the forest slopes, and elephants frequently visit a creek in front of the cabin. But no other animals in the forest leave quite the same trail as gorillas. They trample a yard-wide swath through dense vegetation and leave visible foot and knuckle prints when they emerge onto dirt paths. Bent foliage shows the direction of travel. Signs remain readable up to six or eight weeks if undisturbed, but trails sometimes become obscured when a herd of cattle, elephants, or buffaloes traverses the route.

Buffaloes are abundant on the slopes and saddle areas, even at elevations up to about 12,000 feet. They often use trails that the gorillas make through thick foliage and may follow closely behind a group. Gorillas sometimes find them too persistent and will bluff

Diffident diner mouths strands of Galium vine, a major food. Gorillas also eat wild celery, thistles, nettles, blackberries, tree bark, ferns, and bamboo. Ground foliage is abundant in moist mountain forests. Succulent and dew-laden plants supply water. Recent observations have verified that some gorillas living in the wild vary their diet with insects, snails, and slugs.

BOB CAMPBELL

a charge, giving every appearance of enjoying their ability to send a herd of mighty buffaloes scattering off in all directions. Though the two species share a liking for many types of vegetation, they usually eat different parts of the plant. For example, buffaloes nibble tops of tall nettles while gorillas eat roots, lower stems, and leaves. Forage is plentiful and regenerates quickly, but fruits are sparse.

Gorillas generally feed for about three hours in the morning, rest at midday, then feed and travel on to a night nesting area. The entire group settles down at dusk in crude beds of branches and foliage, shaped into a roughly circular form. Nests are sometimes built in trees in the wooded saddle areas, but on the more open slopes of Visoke they are commonly on the ground. By examining the night nests of known gorilla groups I can count the members, record births, and learn of behavioral changes, such as what happened when a juvenile of one group lost its mother: It began to share the night nest of the group's dominant male.

The gorilla social unit, called the group, is of fairly stable composition. In my study I have observed nine groups, numbering from 5 to 20 members. I chose four for close observation: Groups 4, 5, 8, and 9. One mature male, called a silverback because the hair on a

male gorilla's back turns silver with age, reigns within each group. Subordinate males (other silverbacks and younger blackbacks) may serve as sentries and guards. A group's females, juveniles, and infants enjoy a protected status.

I have found that the character of a group is frequently determined by the character of its leader. In one instance I had an opportunity to see what happened when the leadership changed. When I first encountered Group 4, which included Bravado, it was under the calm rule of a silverback I named Whinny because he was unable to vocalize properly. Whinny died, and leadership fell to the group's second silverback, Uncle Bert. He clamped down on the group's activities like a gouty headmaster. The gorillas' previous calm acceptance of my presence was replaced by chest beating, whacking at foliage, hiding, and similar activities showing alarm. Nervous and excitable in his new role, Uncle Bert led Group 4 higher on Mount Visoke into more remote areas where interference from both humans and other gorillas was minimal.

Until a group becomes well habituated, play seems to be one of the first activities inhibited by the presence of an observer. For this reason, I consider play more common than was once thought. Males and females of all ages play, but playing is most prevalent among infants and juveniles. A popular game is sliding. An infant practices on its mother's body, then (Continued on page 218)

Gallery of gorilla photographs aids Miss Fossey in documenting her daily field notes. Eyeing the California-born scientist from the screen is Uncle Bert. Close-ups of noses, used as noseprints, identify individuals, especially infants, whose color and other features change as they grow. To supplement her notes the author maps ranges and routes, tapes vocalizations, analyzes dung for parasites and type of food consumed, charts nest locations, and records rainfall and daily temperature.

Skulls and bones collected in the highlands aid studies of how the mountain subspecies differs anatomically from gorillas in other African locations.
BOB CAMPBELL

Gorilla see,
gorilla do:
in mother's
arms or in
the group,
infants absorb
habits
and skills

Snug in a sunny arboreal nest, a mother and wide-eyed infant rest after a morning of foraging. The infant, smaller at birth than the average human baby, depends on mother for food, transport, and example. Though suckled until nearly 2 years old, it begins at about three months to learn which plants are good to eat. Imitation or trial and error accounts for most of its newly acquired skills.

The two- to three-hour midday pause in the group's foraging is typically a time of social harmony. Adults and juveniles make simple day nests of bent foliage or branches. Here restless infants romp while mothers nap or groom — picking bits of dried skin or foreign matter from their own or another's hair. Gorillas do not bathe, dislike entering water, and

seek out fallen logs when crossing streams. An infant usually sleeps with its mother until about 3 years old or until she bears again, but it plays at nest-making and may occasionally sleep out nearby.

Adults normally travel on all fours; a mother with a very young infant may clutch it to her chest and walk on three limbs. At five or six months a youngster learns to ride on its mother's back, jockey fashion. True to her name, Old Goat — carrying 2-year-old Tiger — glares at the photographer.

Females of a group benefit by staying together. Their infants, growing more independent, mingle in play, sometimes cared for by an "aunt" — one of the females — while their own mothers feed or sun themselves some distance away.

217

graduates to dirt banks and tree trunks. The favorite time for play sessions seems to be a sunny morning or after a midday rest.

One afternoon I watched as a group reached an open lava slope. Two young adults chased one another across the clearing. Others stopped feeding to watch. Then, for the next 20 minutes or so, the rest of the group joined in, tumbling and rolling across the slide area. At the edge of the clearing, they grabbed branches of giant *Senecios*, swung on them until they broke, and, still holding the foliage, rolled in a jumble down the slope.

Even during play sessions, silverbacks show concern for the safety of their young. The infants and juveniles of Group 5 seem especially secure in the protection of three silverbacks, Brahms, Bartok, and Beethoven. Yet their antics appear to try the patience of even these experienced older animals.

One day Icarus, a wizened, elf-eared little fellow, was trying out an acrobatic routine in a sapling some ten feet away from me when the tree came crashing down, Icarus and all. The noise triggered a somewhat hysterical bluff charge by the silverbacks. They came at me as if my presence had caused the mishap. They halted five to ten feet away when they saw Icarus, none the worse for his spill, calmly climb another tree. But they remained tense, giving frequent barks of alarm. Then to my dismay a small infant climbed into the same broken sapling and began a series of spins, twirls, leg hangs, kicks, and chest pats—all the while exuding blasé self-importance. No high-wire artist ever had such a rapt audience. The eyes of the silverbacks darted back and forth between the infant and me. When our glances met, they roared threateningly.

Surprisingly, it was Icarus who broke the tension. He launched a game of tag with the infant which led them both back to the group. Brahms beat his chest to relieve tension, then ran downhill noisily. Bartok and Beethoven followed suit.

Though my study of the mountain gorilla is not yet complete, one conclusion is already clear. Men malign the gorilla when they portray it as a vicious beast. After more than 3,000 hours of direct observation, I can account for less than five minutes of what might be called "aggressive" behavior. And even this amounted to protective action or bluff. That was the nature, I am sure, of my most

Trekking a high and verdant land, Dian Fossey and two Rwandan assistants end a day of field work and head for camp. Mist swathes Mount Mikeno, extinct volcano of the Virunga chain. From this mossy meadow at 10,200 feet rises Mount Visoke, on whose southwestern slopes dwell gorilla groups (below).

Fluctuations in their ranges, plotted for one-year periods, reflect internal changes in group structure and behavior.

Year one: Group 4's leader and Group 8's lone female both die. Uncle Bert becomes the dominant silverback of Group 4.

Year two: All-male Group 8 seeks proximity of 4 and 9— but overprotective Uncle Bert moves 4 higher on Visoke to avoid contact. Group 8 then concentrates on following 9; Group 5 spreads out eastward.

Year three: Bachelors of 8 continue pursuing 9; Group 4, moving lower, reoccupies much of its former range; 5 expands in saddle area as encroachment of man presses from the east.

▲ Author's camp
Group 9
Group 8
Group 4
Group 5

Year one: As stable units, groups occupy nearly separate ranges.

Year two: Social flux results in near-merger of ranges of 8 and 9.

Year three: Given room to spread, Group 5 flanks the author's camp.

BOB CAMPBELL. DIAGRAMS BY
BETTY CLONINGER, GEOGRAPHIC ART DIVISION

dramatic encounter, in which five large males charged at me, roaring explosively. They stopped—the leader was only three feet away—when I simply spread my arms wide and shouted "Whoa!"

Naturally an animal will try to protect itself, and there are records of gorillas attacking men who hunted them. Typically, males scream or *wraagh* when alarmed, but I have observed that in situations of extreme fear, as when poachers are approaching, a sudden silence effectively alerts the group to danger without pinpointing its location as sounds would.

Based on the distinctiveness of the sounds, I would estimate that gorillas use 16 or 17 different vocal signals. The most common used within a group is what I have named the belch vocalization. Exchanged in situations of maximum contentment, as in the early evening before building nests, the sound resembles a stomach rumble. Usually, when one animal expresses well-being by a belch vocalization, the gorillas nearby respond with similar sounds. On occasion, I have crawled undetected into the midst of a contentedly feeding group, begun my own series of belches, and had them answered in succession by the gorillas around me.

Other vocal signals include pig grunts and hoot barks. The pig grunt, harsh and staccato, is used for discipline, as when a silverback intervenes to settle a squabble or when a female scolds an infant. The hoot bark expresses curiosity or mild alarm. Given by the dominant silverback, it usually alerts the group.

Silverbacks are the most vocal of all the gorillas, and with few exceptions they alone emit and answer the hoot series, a prolonged chain of *hoo hoo hoos*. I have heard this vocal probe only during contacts between two groups or when one silverback of a group is checking on the location of a male traveling apart from any group. The distance between two silverbacks calling each other varies from 20 feet to half a mile or more.

None of my observations suggests that a sense of territoriality exists among gorillas, for I have seen no action resembling defense of a home range. The location of other groups, however, does play a role in regulating a group's movements.

One of my specific aims is to study the reasons for the range trends of the mountain gorilla. Are there consistent relationships between range and food-seeking, range and human intervention, range and the distribution of other gorilla groups? Such knowledge

In play and display the gorilla seeks an outlet for its energy and for pent-up excitement. Chest beating can convey a variety of messages. Rising before an observer, a blackback beats a tattoo as if to say, "Look at me. I'm not afraid of you." A silverback may beat his chest to communicate, as in signaling the location of his group—or to intimidate. An infant hard at play may pat its chest in sheer exuberance.

At the slightest provocation juveniles and young adults square off in strenuous wrestling bouts: head and body holds, mock biting, tumbling, and whacking (below).

Older animals seem to curb their strength when playing with the very young; immune to rules, the infants pull hair, bite, kick, punch, and flail. When the senior has enough, he may simply lean on the young rowdy until it quiets down.

DRAWINGS BY GEORGE FOUNDS

becomes ever more vital as the gorilla domain shrinks. Many groups continue to visit the vicinity of old and familiar forage sites at lower altitudes which have only recently been swallowed up by human settlements. For this reason it is important to know the ranges of the remaining groups so that protection may be concentrated in those areas. And from the point of view of gorilla behavior, the study of range patterns is a fascinating subject.

Human encroachment on Mount Visoke has restricted the apes mostly to the steeper slopes, and here group ranges frequently overlap. When several dominoes are closely aligned and the first is pushed, all are affected. I feel almost certain that the same concept can be applied to what is happening to the ranges of my four study groups. During the past four years a change by one group, almost imperceptible at first but growing dramatic as it progressed, has vastly affected the range concentrations of the other three groups. The leading domino was Group 8.

Koko, the lone female of Group 8, died. An elderly, doddering female with atrophied arms, dried-up breasts, and graying head,

Koko had a personality that exerted a cohesive influence on the five males. Despite her advanced age, which I estimated at about 50 years, she was able to induce play activities and mutual grooming. When she started the grooming—a kind of social behavior involving meticulous hair parting, searching for and plucking particles —the others would do it too. Within a few minutes there would be an entire chain of intently grooming gorillas. Group activities centered about this aged matriarch.

But after her death, Group 8, headed by Rafiki, the dominant silverback, tended more and more to seek its neighbors, Groups 4 and 9. Initially, Rafiki and his fellow bachelors were not tolerated by either group. Hours of parallel displays by the silverbacks—long hoot series, chest beating, ground thumping, and runs through the foliage—usually marked an encounter.

Rebuked, the five males continued with bulldog persistence, following first one group and then the other, though intermingling was rare. Not content with their own company, the Group 8 gorillas would travel up to two days in direct *(Continued on page 228)*

A touch of warmth: primate meets primate in an African forest

One drizzly morning I set out from camp on Mount Visoke to contact one of the groups that I have studied closely for the past few years.

As I was about to descend into a gully between two open slopes—a popular gorilla gathering place—I heard a soft hoot bark coming from above. Quickly I climbed into the crotch of a large *Hagenia* tree, settling as inconspicuously as possible. Through the foliage I saw two brawny black arms encircle the trunk of a tree slightly higher on the slope. Then two bright eyes peered from a lattice of ferns. It was Peanuts of Group 8, one of my favorite gorilla personalities.

Today he was wearing the "fun and games" expression that showed he wanted to prolong a contact with me. Peanuts appeared so relaxed and interested that I slowly climbed down into the ground foliage and began to make feeding noises. The fleeting moments that followed are among the most memorable of my life.

After sizing up the situation, the young blackback came down from the tree for a bit of strutting and swaggering. A flawless showman, he varied his act by throwing leaves, beating his chest, and as a finale strutting and slapping at foliage.

Suddenly he was at my side. He sat down, eyeing me as if to say, "I entertained you. Now it's your turn." Evidently he did not consider my feeding, baby talk, or singing to be much of an act. Not wanting to lose his interest, I began to scratch and groom myself. At last he responded. Peanuts too began to scratch, watching me as intently as I was watching him. It was impossible to say who was aping whom.

Until that day the part of me that is all scientist had ruled against my reaching out to stroke a wild subject, though I have often wanted to. But now the acute rapport I felt with Peanuts made an overt gesture on my part seem appropriate. Lying back among the leaves, to appear as harmless as possible, I extended my palm. Then slowly I turned my hand over and let it lie.

To say of a wild creature that he "pondered" may offend some scientists, but Peanuts did seem to ponder this familiar yet strange object

BOB CAMPBELL

224

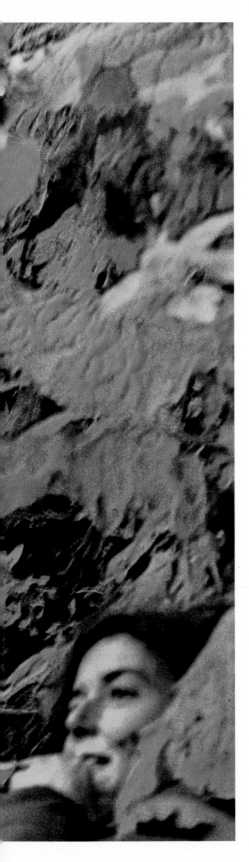

extended toward him. He stood up, hesitated, and began to edge closer, all the while looking at my hand. I believed that he meant to do more than just look. My excitement mounted. My face was one immense grin; I worried that this might deter him and put up my left hand to partly cover my mouth.

What seemed an interminable time went by as Peanuts put out his furry hand and gently touched his fingers twice against mine.

In actual time, it was over so quickly that Bob Campbell, whose excellent photography has proved so valuable in documenting my studies, clicked the camera shutter an instant too late. He caught Peanuts in the process of retracting his hand and me expressing incredible joy—by crying.

For that fleeting instant a bridge, spanning a chasm of immeasurable time, linked our two species. I thought myself the most fortunate person in the world to have gained the trust of this wild animal, though for him the touch probably meant no more than a chance to satisfy curiosity—part of the fun and games.

After he withdrew his hand, Peanuts sat down. I left my hand exactly where it was, but he did not venture to touch it again. He stood up, gave vent to his suppressed excitement with a whirling chest beat, and ambled off to rejoin Group 8, feeding nearby. I don't know whether he looked back, but I doubt it.

pursuit, thus extending their known range significantly. Obviously, bachelorhood wasn't all it was cut out to be.

Group 4, having recently changed leaders, had troubles of its own. When insecure Uncle Bert led his group higher on Mount Visoke, the "terrible five" concentrated on following Group 9, which fled farther and farther north. Sometimes Geronimo, the leader of Group 9, would keep its members on the move for hours with little or no time out for feeding or day naps. They often succeeded in losing Rafiki's group, but only for a few days at a time.

Group 5, meanwhile, maintained a relatively stable range for almost two years, rarely encountering the other three. Then this group, not at all shy with its three silverbacks, began to make probes into the saddle area near my camp.

During the third year this pattern emerged: Group 4 (as Uncle Bert became more outgoing) cautiously returned to the slopes it had previously ranged; 5 expanded its toehold near my camp; 9 continued slowly north but with increasing tolerance of 8.

A new trauma suddenly reshuffled my study groups. Group 9 was cut from 17 members to five. Perhaps this drastic reduction in the group they had been following was the reason Rafiki and his friends turned their attentions back to Group 4.

I could only speculate about what happened in Group 9, which is seldom seen in my study area now. The 12 missing members may have split off and merged with another group on the north side of the mountain. Or they may have been killed. Not long before, there had been a gorilla slaughter. The bodies of five gorillas—members of a fringe group that occasionally visited the foothills—had been found in an area south of my camp. They had been mauled by dogs, pierced by spears, and battered by stones, apparently just for the excitement of the hunt. Their fate gave me real reason to worry about my friends of Group 9.

The return south of Rafiki and his bachelors, meanwhile, proved an event with vast behavioral implications. Within a week, to Rafiki's obvious delight, Group 8 had acquired from Group 4 a female I knew as Maisie, along with a blackback named Macho.

For a while Group 4 followed Group 8 closely, as if hoping to reclaim Maisie and Macho. But then Group 4 added its own new member when Flossie gave birth to little Cleo. Uncle Bert then seemed quite content overprotecting the new addition.

Despite his attentions to Maisie, Rafiki soon lost her to a younger silverback. Maisie, fickle female, was seen traveling in the company of Samson, a Group 8 dropout. And once again the bachelors were left alone, shadowy figures on a dwindling domain.

Shy kings of the forest file up a steep mountainside, a scene suggestive of their uphill flight for room to live. Victims of human folly fed on myth and misunderstanding, so many have been killed that conservationists now class mountain gorillas as "rare."

Even in refuges like Rwanda's Parc des Volcans, scofflaws still molest these great apes, and the land hunger of herdsman and farmer forces them toward peaks where sparse vegetation cannot fill the gorillas' needs.
BOB CAMPBELL

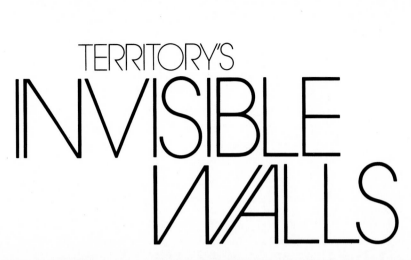

Richard D. Estes

TERRITORY'S
INVISIBLE
WALLS

n a windswept hill near the eastern side of Ngorongoro Crater stands a wildebeest bull. He is alone, a solitary figure knee-deep in a sea of billowing red oat grass. Behind him looms a steep and weatherworn escarpment — the rim of a collapsed volcano that rises above the nearby Serengeti Plain of Tanzania.

Below him, spread across the ten-mile-wide crater floor, 15,000 wildebeests cluster in a great number of small herds and several large aggregations. They share the lush grasslands with 10,000 gazelles, zebras, ox-like elands, and other herbivores, and are in turn a source of food for the carnivores: resident lions and spotted hyenas, as well as visiting cheetahs and wild dogs.

You will find the solitary bull at his outpost day after day, per-haps month after month — if nothing happens to him. But what

Untouched by the territorial urge, young wildebeests gambol across East Africa's Serengeti Plain.

Separated from their mothers as yearlings, males live two or three years in bachelor herds. Then maturity awakens their drive for territory, impelling each to seek a place of his own. For only if he holds a plot of land can a bull ever hope to mate.

Wildebeests are classed among the continent's 70 or so antelope species. Bulls stand more than four feet high at the withers and weigh up to 550 pounds.

BRUCE DALE, NATIONAL GEOGRAPHIC PHOTOGRAPHER

*Migrating wildebeests stream
across the Serengeti on trails
etched by centuries of hoofs.
The treks crest spectacularly in
May and November when a million
or more zebras, wildebeests,
and gazelles sweep over the plain
in tides that flow as far as
the eye can see. Greening grass
triggers the winding 800-mile
march along a triangular course
into differing ecological areas.*

*Wildebeests remain territorial
even during migration. Wherever
they pause to graze or rest,
bulls quickly stake out claims.
They guard their property
against other bulls, break up
large groups of cows and calves,
and shunt landless bachelors
to the sidelines. When the herds
move on, the bulls hasten
to overtake them and set up new
territories at the next stop.*

*The wildebeests of Ngorongoro,
unlike their wandering kin
on the Serengeti, seldom venture
outside the walls of their
crater paradise. Those that do
leave do not go far and usually
return in May, when rainwater
pools on the plain evaporate.*

THOMAS NEBBIA

232

brought him to this lonely place? And what keeps him so far removed from the companionship of the peacefully grazing herds?

Such questions occupied my thoughts when I began a 2½-year study of the behavior of wildebeests and other ungulates that inhabit the crater. I would find the answers only after thousands of hours observing these hoofed animals in their natural surroundings.

High-shouldered and homely, the gregarious wildebeest—or gnu as it is also called—is an antelope closely related to the topi and the hartebeest. Its antic behavior, especially when disturbed, has earned it renown as "the clown of the veld."

"As one approaches a herd of wildebeest," wrote Theodore Roosevelt, "they stand and gaze, sometimes snorting or grunting, their heads held high and their manes giving them a leonine look. The bulls may pace up and down, pawing the ground and lashing their tails. Then, as the hunter comes nearer, down go their heads and off they start, rollicking, plunging, and bucking. . . . or a bull will almost stand on its head to toss up the dust with its horns. . . ."

This behavior, amusing as it may seem, is not motivated by high spirits. It is what happens when an intruder—man or beast—disrupts a territorial network, forcing bulls to flee through property vigorously guarded by their neighbors. The wholesale violation of boundaries sets off a chain reaction of furious activity.

The wildebeest on the hill is not a sick animal, nor is he an old bull cast out from the herds. He is, rather, a territorial male in his prime. He waits as if anchored to his property on the chance that females will wander through it. Since the cows must come to him, he must have a place to call his own before he can have the opportunity to mate. If the bull's vigil proves fruitless, he eventually abandons his post, although in so doing he forfeits the one place where he holds sway over all other wildebeests. And if he fails to win himself another territory—preferably one closer to the nursery herds of cows and their young—he will end his days in a bachelor herd, in the company of bulls that have exchanged the chance to breed for peace and companionship.

It is the males in these bachelor herds—adolescent bulls biding their time to full maturity and old or weak males unequal to the task of gaining or holding territories—that are the true outcasts of the wildebeest social system. Driven to the fringes of the territorial network which their more belligerent brothers establish wherever nursery herds are found, they lead celibate lives. And often they are forced into areas of high grass where predators may lie concealed.

The size of a bull's territory varies greatly, depending on location, season, and the presence or absence (Continued on page 238) 233

The challenge ritual: territory's imperative test

Territory is the measure of the wildebeest bull. Without it he is ignored by females — and his land-owning brothers chase him to the outermost pastures of the habitat. With it, he is master of all the wildebeests that enter his realm. But to maintain his position he must continually prove himself through a form of psychological warfare known as the challenge ritual. These highly stylized skirmishes seldom result in bloodshed, last from 1 to 15 minutes, and average a good 45 minutes of a bull's day.

"The initial provocation is normally a deliberate act of trespass," Dr. Estes writes, after recording 100 duels. This photographic portrayal of an unusually combative encounter is based on dozens of clashes filmed by the author in Ngorongoro. As taken from field notes: "April 29, 11:35. I [Invader] enters H's [Home Bull's] territory, walks past and stops a little behind him. . . ." Home Bull (below, background) presents his flank to Invader in a challenge stance called the lateral display.

The two antagonists stand grazing head-to-head, warily taking one another's measure (center).

Suddenly, the bulls whirl and drop to their knees (upper), lashing their tails and breathing defiance as they graze eyeball-to-eyeball in the combat attitude. Neither quite dares launch an attack.

Heads high in a mock alarm display (lower right), the snorting animals apparently ease tension by pretending to sense danger, such as a predator.

RICHARD D. ESTES

Cavorting (left), another spectacular alternative to fighting, often occurs when an invader refuses to take part in a ritual. The landholder shakes his head violently and soon works himself into a frenzy of spinning, bucking, kicking, and leaping.

Fights, though rare, do occur. Bulls brawl on their knees, ramming and swinging their horns like duelists. Most fights end after a few shoves.

Peace restored (below), a bull with head proudly raised starts toward a herd of cows and calves that have wandered onto his property.

Challenge rituals can involve up to 30 distinct actions performed by either or both contestants in no particular order. No two bouts are exactly alike. Challenges, along with such territorial displays as scent-marking, pawing, or horning the dirt, are common to many antelopes. But two seem unique to wildebeests in their territorial encounters: the head-rump rub, when opponents mark one another's haunches with scent from facial glands; and an olfactory check of territorial credentials through urinalysis. Called *flehmen,* the performance is normally elicited only by the odor of female urine. Wildebeest bulls also seem to use it to assess each other's hormone level. (Territorial males may produce more testosterone than landless ones.)

On demand, a bull assumes a show-horse stance and urinates. His opponent puts his nose in the stream and lifts his chin in a gum-exposing grimace. Standing transfixed nearly 30 seconds, the challenger opens and closes his nostrils, rapidly licks his gums—then often urinates in turn.

of females. During the peak rutting weeks at the end of the East African rainy season, when about 80 percent of the cows come into estrus and are bred, bulls in the midst of large concentrations may be only 30 yards apart. Where herds are few and far between, holdings are correspondingly larger: Males occupying territories near the edge of the network may be half a mile or more from each other.

A bull's plot of ground is irregularly shaped, its boundaries invisible. Apart from the owner's presence, the only sign of ownership is a dung-strewn bare spot the size of a tabletop. This is the bull's stamping ground, heart of his realm.

Even though a territory appears indistinguishable from the surrounding featureless plain, it is a definite piece of real estate, every inch known to the owner. An observer can determine its limits only by plotting the positions occupied by a bull over a period of time. As fixed reference points, I first tried numbered stakes, which zebras knocked over and hyenas chewed up. Then I hit on the solution of using large whitewashed stones.

A homesteading bull who seeks to establish himself close to the center of a territorial network faces seemingly insurmountable obstacles. Imagine a neighborhood where all the residents are

downright hostile, where a newcomer is tolerated only if he is prepared to stand up to the owner of every bordering lot. Add to this the
fact that until he has won a place of his own the newcomer has
little stomach for fighting, and you have a fair idea of the dilemma
that faces the homesteading wildebeest. But with motivation, luck,
and above all, persistence, he can win out against the odds.

He first probes for a weakly defended spot—not hard to find if
he is content with a substandard location near the edge of the network. But a vigorous bull tries to get as close as possible to the center
of the action, where real estate is most zealously defended. Here
bulls have more opportunity to breed; thus natural selection favors
them over their less ambitious brothers.

The prospecting bull, chancing upon a place where he is un-
molested, lingers until challenged by a landholder—whereupon he
runs away. But as soon as the established bull's back is turned,
the homesteader comes back—again and again, day after day, until
at last the surrounding property holders become so accustomed to
his presence that they seldom bother to chase him.

The longer the newcomer is allowed to remain, the more confident he becomes. As his courage swells, he begins to advertise his

new status, calling at and bullying passing wildebeests and creating his own stamping ground by pawing, dunging, rolling, and digging his horns into the dirt. These activities reinforce his sense of ownership until, finally, he is prepared to stand up to his neighbors and they are forced to treat him as an equal.

But even on the brink of acceptance he may fail. I have seen an apparently secure newcomer suddenly lose nerve and run away as a neighbor, intent on chasing cows, entered the newly won territory. When this happens, the newcomer must start all over again.

Male and female wildebeests reach adolescence as yearlings, at which time heifers can breed. But few young males can meet the rigorous tests of territoriality until they are nearly 3½ years old, when they reach adult weight and full sexual development. At the same time, increased hormonal activity associated with the rutting season probably raises their aggressiveness to the threshold needed to leave the bachelor herds and stake out a territory. Even so, many adult bulls fail to make the grade. Of several thousand in Ngorongoro, only about half are territorial at any one time. And even during the excitement of the rut many fit-looking bulls can be seen in the bachelor herds.

Both sex and aggression, the two main ingredients of the territorial drive, are fueled by the male hormone, testosterone. But the full expression of male drives is utterly dependent on possession of property. What is most amazing about territoriality is that it is not only physiological but also psychological—as proved by the fact that a territory owner switches off virtually all displays of sex and aggression whenever he leaves his property.

Females apparently have no territorial urge. Most of them roam widely throughout the crater in large loose-knit groups numbering a hundred to a thousand or more animals. Others, usually led by an old matriarch, form semi-permanent nursery herds averaging ten cows and calves. They tend to stay within a limited range—often less than a square mile—while short succulent grass is plentiful.

In Ngorongoro, where there is a herd for every five to ten territorial males, no bull can keep the cows on his property for long, especially during the peak mating weeks of May and June. Even a bull that manages to establish himself in a central location still passes most of the time in solitude. He spends his days simply grazing, chewing his cud, or standing relaxed and motionless on his stamping ground, head raised in a distinctive pose. From time to time he paws the dirt vigorously, kneels and rubs his face on the ground, or rolls on his back, kicking up a cloud of dust in the process. If he sees a nursery herd cantering across the plain, he and his neighbors break into a chorus of deep grunts—rather like the croaking of giant frogs. But he will not leave his land, and if the herd passes him by, he will go back to grazing.

Individually, such activities may appear meaningless. But by repeating and combining many of them as ostentatiously as possible, a bull unmistakably signals his territoriality. The grunting call is given only by bulls; cows and calves moo. Rubbing the head on the ground scent-marks the territory with musky secretions from glands located near the eyes. Kneeling and horning are visual displays of marking and aggression, as is defecation when preceded by vigorous pawing. Glands between the bull's cloven hoofs also imbue his stamping ground with scent whenever he paws. Rolling,

Silhouetted against the Serengeti sky, a topi boldly advertises his presence to females grazing on the horizon. These antelope relatives of the wildebeest often seek height not only to see but to be seen. They can hold their pose for hours—even while dozing. Here, atop a termite hill, the animal's proud posture proclaims his territorial status. Such "static-optic" displays, usually on a more modest scale, are basic to most ungulates. Among the continent's antelopes, only spiral-horned species such as bushbucks and elands apparently hold no territory.

GEORGE B. SCHALLER

unknown in other antelopes, saturates the owner's coat with his various scents and probably also doubles as a visual display, because only males behaving territorially do it. Even standing with head raised advertises a bull's territorial status, for cows and bachelors almost always carry their heads at or below shoulder level. The closer wildebeest bulls are to one another, the more frequent and emphatic their displays.

A bull approached by cows and calves moves to intercept them in a characteristic rocking canter, head raised and grunting every second or two. If they enter his domain, he either chases them out or tries to keep them in, depending on his mood and how the cows behave. He signals sexual interest by lifting his chin and holding his ears at half-mast. If he is aggressive, he shakes his head, lashes his tail, and chases the intruders away.

Trespassing bachelor herds let themselves be evicted with as little resistance as a herd of cows. A lone bull, often a territorial male on his way to or from a water hole, is rarely belligerent but will usually defend himself rather than be put to flight. He is chased only if he fails to stand his ground.

As the transient bull comes near, the owner straddles his path as though disputing the intruder's right to pass. To avoid a confrontation, the visitor simply alters his course to pass behind the landholder and continues on his way. To pass in front, presenting his side to the owner, is to invite attack.

By far the most elaborate behavior in the repertoire of the wildebeest is displayed in the so-called challenge ritual between territorial bulls, an intricate series of actions and counteractions by which bulls probe one another's territorial fitness without undue risk of bloodshed (page 234). A challenge ritual lasts an average of six or seven minutes and is commonly provoked by a deliberate act of trespass—as when a bull saunters onto his neighbor's stamping ground to defecate or to horn the dirt. Through these encounters a bull's defenses are tested daily by each of his neighbors in a game of bluff and counterbluff that largely substitutes displays of combat readiness for actual fighting.

Fighting, the ultimate test of fitness, may break out in a high-key duel, but only rarely does it amount to more than a brief joust and even more rarely does it result in serious injury. Marathon bouts lasting up to half an hour do occur, but usually only during the excitement of the rut or early in the rainy season, when many bulls contend for pastures that have been abandoned. If a contestant in such a fight finds himself outmatched, his only recourse is to run away. He thereby saves his hide but loses his land. Getting it back

Lips curled, head raised high, a sable antelope assumes the flehmen posture amid milling females. He may seek a mate by chemical testing: Specialized olfactory receptors adjacent to the nasal passages are believed by the author to assay hormone concentrations in urine. Called the vomeronasal, or Jacobson's, organ, the tiny cigar-shaped structure may enable males of many mammal species to gauge the onset of estrus in females.

A Thomson's gazelle urinates in the show-horse stance; he will then defecate on the same spot. The linked ritual scent-marks his property and is also a visual show of ownership performed by territorial males exclusively.

Another Tommy twists his head to mark a blade of grass with a pitchlike secretion from facial glands. Such scent signposts, placed throughout a territory, warn other males not to trespass. The markings may also bolster the owner's sense of possession.
RICHARD D. ESTES

243

may be harder for him than for an outsider trying for the first time. Former neighbors actually seem to go out of their way to keep a previous landowner from regaining his territory.

The first bull I immobilized and branded lost his territory within two weeks of the incident. Thereafter, he was seen in a large bachelor herd two miles away. Although his former place was promptly claimed by his neighbors, I saw him in the immediate vicinity 19 times over the next six months. Twice he was seen fighting and once in a challenge ritual. During the following year he occupied land close to his old territory three times—only to lose it in a matter of days. On a visit to Ngorongoro after an absence of four years, I was gratified to see that he had regained his old place.

Antelope territoriality serves reproduction by limiting sexual competition to property holders and by forcing males to wait in place for females to come to them, rather than having them cluster around the cows in a grand free-for-all. But the territorial system serves other useful purposes as well. The bull's intolerance of too many cows and calves on his property helps break large concentrations into small groups, thus minimizing damage to the pasture by trampling and close grazing. Forcing bachelors to the fringes of the habitat reduces the confusion during the calving season. Expulsion of the bachelors also lessens competition with cows and calves for choice pastures.

The territorial network amounts to an early-warning system, with sentinels posted around the nursery herds. When a bull sees a predator skulking through the grass, he signals with a head-high alarm posture and a snort—both universal warnings among plains game. This immediately alerts all nearby animals to the danger and allows them a few extra moments to flee.

If there is one imperative for a territorial male, it is the self-assurance stemming from the feeling—call it conviction—of superiority to everyone else on his property. Away from that property, the bull loses this conviction. His aggressive drive recedes and his suppressed gregarious drive becomes dominant.

In the dry season, when good grazing is reduced, many territorial bulls commute to remaining green pastures or form temporary bachelor "clubs" around water holes. Others may sojourn with bachelor herds. But some bulls are so attached to their property that they cannot bear to leave it—even when all other wildebeests change their range for another part of the crater. So they tarry, and sometimes an animal will watch his companions drift away while he remains, rooted to his patch of land—like the lone bull in the oat grass near the eastern side of Ngorongoro.

Hunter and prey meet in uneasy truce on the Ngorongoro plains. This lion, detected in time by his intended victim, poses no threat—at least for the moment. But territorial bulls are not always so lucky. Those that hold exposed positions where herds seldom gather make tempting targets for prowling carnivores, especially at night when cows and calves bed down in close formations. Despite the hazards, landholders seldom leave their property except for a drink or to visit greener pastures in times of drought. The author, in more than 1,200 visits to territories held by known bulls, found an average of 90 percent of the owners present. With luck, a wildebeest may reign over a domain for as long as a decade.
RICHARD D. ESTES

THE
WARRING CLANS
OF THE
HYENA

Hans Kruuk

Snarls and screams filled the moonlit August night as the two hyena packs lunged at one another. Hackles bristling, fangs snapping, hyenas seemed to be everywhere, swirling around the Land-Rover in a dizzying storm of sound and fury. The carcass of a wildebeest, freshly killed by one of the packs, lay momentarily abandoned in the pandemonium. I struggled desperately to report the event into a recorder, but the effort left me breathless and I broke off with a final entry: "No use trying."

My wife Jane and I had come to Tanzania's vast Ngorongoro Crater to study the spotted, or "laughing," hyena, the most common of Africa's three known species. Day and night during the months of our stay we made thousands of observations. Drug-laden darts fired from a gun enabled us to immobilize 50 hyenas for brief periods so

Eyes ignited by a camera flash, spotted hyenas tear into a Burchell's zebra in Tanzania's Ngorongoro Crater. Massive bone-crushing jaws can demolish a carcass in less than an hour.

The author, a young Dutch scientist, found that hyenas, long scorned as mere scavengers, also are bold hunters that roam in packs by night. Preying mostly on wildebeests and zebras, they chase down their victim and kill it by disembowelment.

Hyenas lope long distances almost tirelessly and will plunge into water in pursuit of quarry. Sprinting 40 miles an hour, a hyena can overtake all but the swiftest animals.
ALAN ROOT

that we could examine them and notch their ears for later identification. Gradually we pieced together an entirely new picture of
their relationship to other animals in the crater.

At first Jane and I, like most people, thought of hyenas as scavengers dependent on kills made by braver animals. But after watching
the packs in action, we realized they are also highly efficient predators, living mostly on wildebeests and zebras. And their hunting
skills serve not only themselves, as we saw in dozens of encounters,
but also the lions. Ngorongoro's lions, unlike those elsewhere, seldom hunt. Instead, they plunder hyena kills, guided to the feast by
the eerie "laughter" of the feeding pack. Hyenas will abandon a kill
to several marauders, but often defend it against a solitary intruder.
Once we watched a whooping mob drive a lioness off their kill,
snapping at her rump until she fled, growling, into the bush.

As our study progressed, Jane and I learned that Ngorongoro's
hyenas live in female-dominated clans of up to 80 animals.
We further learned that the clans have divided the 100-square-mile
bowl into eight territories and that each group usually stays well
within its own area. Boundaries between clans, while not defined
by natural features such as streams and hills, are regularly patrolled
and marked with feces and scent from anal glands. And the boundary lines are clearly marked in hyena memory.

What happens when one clan violates a neighbor's territory? We
found the answer this moonlit August night. My wife and I had set
out after dinner, following a hunting pack from a group we named
the Clan of the Scratching Rocks because their den lay close to
some boulders that zebras liked to rub against. And now, as we
watched the successful hunters gorge on the wildebeest cow, we
were startled by deep, staccato grunts—the hyena alarm call.

The animals sprang from the carcass, staring into the night. Then
we too saw the newcomers materializing from the shadows, rushing headlong toward the hunters. First there were four strangers,
then five, then six. Finally there were about 30 hyenas in two small
armies. Members of each group crowded together, snarling and
"giggling," tails aggressively bent forward over their backs.

One old female from the Scratching Rocks Clan took courage and
approached the carcass. Immediately three opponents attacked,
biting a hind leg, forcing her into a crouch to protect her haunches.
And with that, the fight was on.

"There, among the newcomers," I said, pointing, "the one with
the marks at the top of its left ear and the center of its right." Jane
snapped on her light and quickly thumbed through her identification
notebook. "An old female," she said, "from the Lakeside Clan."

Brazen hunters stalk a tempting meal—a month-old rhino calf whose mother removed it from sheltering thickets. The mother, bedecked with tick-eating oxpeckers, appears oblivious to the threat. Hyenas harassed the pair three weeks, giving up only when the infant grew too big to kill. The young or lame of any species—including lions—may fall victim to their attack.

An adult male, meaty prize clenched in his teeth, flees a pack of boisterous cubs. His low status makes him the butt of their playful assaults. Within the hierarchy of a hyena clan, the biggest and boldest animals seize the choicest parts of a kill. Low-ranking older members and newly weaned young often must be content with scraps.

FROM "INNOCENT KILLERS"
BY HUGO AND JANE VAN LAWICK-GOODALL

In a flash we understood: The Scratching Rocks Clan, in the heat of the chase, had downed their prey on their neighbors' territory. Now they were under attack for trespassing.

The fighting raged 15 minutes or so, with first one side and then the other uppermost. Gradually the defending Lakesiders prevailed and the Scratching Rocks hyenas, licking their wounds, slowly drifted back to the safety of their own grounds.

On a cold gray morning several weeks later, Jane and I again followed a pack of Scratching Rocks hunters. A handsome male we called Butcher suddenly sprinted into the middle of a drowsy wildebeest herd, scattering it in all directions. Within seconds Butcher singled out a bull, and other hyenas, anticipating a meal, joined the chase. With four hunters nipping at his heels, the bull appeared doomed. But the hyenas abruptly slowed, then stopped—they had reached the limits of their territory. Though not a single hyena from the neighboring clan was anywhere to be seen, the hunters, perhaps with the memory of their recent battle still fresh, dared not follow through with the hunt.

Looking through my field notes later, I found that fully 20 percent of unsuccessful wildebeest hunts could be attributed to hyena respect for one another's boundaries. But why should Ngorongoro's hyenas be so territory-conscious? The answer, I strongly suspect, is that without such boundaries too many adults would gather to devour a kill, thereby denying food to newly weaned and less adept clan members. It would seem, therefore, that the hazards of territoriality are a small price to pay for survival in the crater.

Trophy of the hunt, the tail of a wildebeest dangles from a young hyena's jaws. Hair and horns are often all that remain after a feast. Cubs sometimes play with them for days.

Snarls and bared fangs drive a vulture from another feast, already picked to the bone. Hyenas share with vultures— but only after satiating their own voracious appetites. Bird and beast may help each other find the next meal. Hyenas sometimes watch for vultures alighting on a carcass; when airborne, vultures can see hyenas prowling for food. Lions also use hyenas as dining guides, homing in on hyenas' hysterical "laughter," an eating signal. The author lured lions to a loudspeaker that blared the recorded sounds of hyena packs gathered around kills.

HUGO VAN LAWICK. OPPOSITE: JEFF SCHOFFERN, FROM "INNOCENT KILLERS"

M. Philip Kahl

THE STORK:
A TASTE FOR
SURVIVAL

They were ravenously hungry. And there were fresh fish and lean raw beef before them, food aplenty for the four African openbill stork nestlings that were sharing my quarters in Kampala, Uganda. Yet they wouldn't touch a morsel.

One of the four had entered my home as an egg. No parent had taught it that African openbills eat little else than snails and mussels. But whenever I could obtain a supply of these, the four nestlings snatched them up from the nest-box floor in a few seconds. One ate its own weight in snails in 12 hours; later it gobbled up a third of its body weight in 20 seconds—and begged for more!

When the snail supply ran low, though, the nestlings rocked back on their knobby heels and spurned nearly every substitute I could offer. When I camouflaged the beef or fish by covering it with

Lanky legs outstretched, a painted stork settles among its comrades in a nesting tree near Bharatpur, India. Sunset silhouettes these Asian fish-eaters; in daylight they show the black-and-white checkering on breast and wings— and in breeding season the brilliant pink covert feathers— that earned them their name.

The 17 species of storks inhabit mainly the warmer regions of the world. Though their ranges may overlap, their tastes vary. In Africa the author watched seven species share adjacent feeding grounds in peaceful coexistence. Thus do the species survive, each in its own "food niche," dining on fare its cousins disdain.
M. PHILIP KAHL

A family tree for scores of
Asian openbill storks spreads
burdened boughs near Bangkok,
Thailand. Both parents feed the
young; when they fetch a feast,
a nestling will gobble up far
more than its belly can hold,
stowing the overflow in its
highly elastic esophagus until
digestion makes room below.

Home delivery brings a meal
of regurgitated tidbits—fish,
eels, perhaps an earthworm or
a frog—to month-old maguari
storks (right) in wetlands of
Argentina. Of all storks, only
maguaris nest on the ground.
Black down may help hide the
nestlings from birds of prey.
The author found young
maguaris difficult to spot
from a low-flying plane.

the bodies of snails, they often sorted out the snails and left the rest. Even liver failed to tempt them; though it looks and feels much like fresh snail, they seldom ate any of it. Finally I had to resort to force-feeding, pushing fish and beef down their gullets at regular intervals. To no avail. Sadly, three of the four starved to death at the banquet table in only ten days.

Ironically, what killed these openbills is what keeps the various species of storks alive: a rigid insistence on a certain kind of food. For an axiom of biology states that no two species with the same ecological requirements can continue to coexist in the same habitat; eventually one must eliminate the other through competition. Yet in East Africa, stronghold of the storks, I have seen as many as seven closely related species feeding within sight of each other. How then can the seven coexist?

The answer is that they do not compete. They share the same habitat but use different food resources within it. While the saddlebill stork watches for fish perhaps six inches long or better, the yellowbill fishes by feel for smaller ones. The openbill forages for snails and mussels, the Abdim's stork for insects. The marabou is mainly a scavenger. The white and the woolly-neck will eat almost anything. Yet even here there is little competition, for the white stork is a migrant visitor in Africa, and the woolly-neck breeds well away from the wandering flocks of whites. By their varied manner of exploiting different food resources, the stork species have effectively sidestepped the problem of competition.

Branching out into diverse niches is common in biology. Evolutionists call it adaptive radiation—that is, the species have radiated out, like spokes in a wheel, from a common ancestor into a variety of niches. Thus each species avoids competing with near-relatives by pursuing a specialized life-style of its own.

Some storks have developed food preferences and feeding techniques that, while enabling them to exploit a certain type of food fully, limit their ability to switch to other fare. Evolution has fitted the adult openbill, for example, with a unique bill structure for dealing with snails and other shelled mollusks. There is a gap between the halves of the bill when the tips are closed.

Until recently it was believed that the openbill commonly used its mandibles as a nutcracker. Not so; the bird usually gains access to a snail by deft manipulation. It manages to insert the tip of the bill, cut the snail's muscle, and extract the snail body in 10 to 20 seconds, often without so much as cracking the shell! At times, even the most proficient openbill finds a snail or mussel that is too tough or tightly closed to extract. Then the bird may crack the shell or leave it basking in the hot sun on shore until the animal within gives up and "opens the hatch" far enough for the stork to wedge its narrow bill inside.

Even in the richest of marshes snails are hard to find. Openbills must keep constantly on the move, peering and probing in the underwater vegetation for their elusive prey. Only the lucky few find an animated plow—the hippopotamus—which not only provides a ride but also turns up the snails for them.

In the Queen Elizabeth National Park of western Uganda, both hippos and openbill storks are common. During the day, hippos retire to wallows, water holes kept deep and muddy by churning masses of hippo flesh. The droppings of these gargantuan residents help support in the wallows a luxuriant growth of vegetation and the snails that feed on it.

Openbills in the area have learned that when the hippo moves, it plows up the vegetation and exposes snails clinging to the undersides. Hippos are strict vegetarians and take no interest in the birds atop their broad backs. The birds seem to know this and ride worry-free, grabbing snails to left and right with very little effort.

Africa's yellowbilled stork, like its cousins in Asia and America, is a fisher *par excellence*. It often forages in muddy or weedy water where it cannot see what it's doing. Birds such as herons that fish by sight must wait for the fish to surface or search in clearer water. But not the yellowbill, which locates its food by feel. It gropes in the water with its opened bill; when the bill contacts a fish, it snaps shut like a mousetrap.

To test this reflex, a colleague and I experimented with the closely related American wood stork. Blackened segments of Ping-Pong balls were glued over the eyes of one bird, which was then allowed to forage with an unblinded companion in a wading pool full of minnows. As they swept their opened bills about, neither seemed to orient visually to the fish, even though the water was clear and the fish were in plain view to the unblinded bird. But whenever a fish swam between the open tips of a bill, the mandibles closed in a flash and the bird came up with a wriggling meal. Blindness seemed no handicap to the one bird. In fact, it caught more fish than its companion, which was more easily distracted by my filming the event from nearby. In another experiment we used an oscilloscope to

Wings flap, bills clatter, and a choice nesting site changes tenants as male yellowbilled storks open the breeding season in Kenya. A black-headed heron watches from ringside. Here the winner will begin building a flimsy nest and—yesteryear's bonds forgotten—select a new mate to finish it with sticks he gathers. Both parents warm the eggs and feed the young.

M. PHILIP KAHL

258

measure the delay between something touching the bird's bill and the bill snapping shut. The average delay was only 1/40th of a second — no longer than it takes you to react with a blink when a loud noise startles you. A fish doesn't stand much chance!

Fishing by feel has obvious advantages as long as the fish are plentiful. When the fish in a marsh or pond thin out, however, these birds have trouble finding enough to eat. Consequently, they must range far afield each day, searching out areas where food is abundant. They often feed in large flocks that exhaust the food supply in an area within a short time, and then move on in search of other ponds of plenty. For the survival of their species a conspicuous plumage pattern of black and white is probably advantageous, since distant companions can more easily spot feeding gatherings from the air and converge to share the feast.

The entire life cycle of the wood stork and its close cousins is so tied to the abundance of food that their breeding behavior is timed to coincide with it. Most other North American birds apparently have their physiological "clocks" geared to the length of the daylight period and thus breed in the spring. But in Florida I found that the breeding of the wood stork seems to be triggered by the increased food density that accompanies falling water levels at the beginning of the dry season in early winter.

South Florida — major breeding ground of the wood stork in the United States — is an extremely flat area where falling water levels trap fish in countless scattered pools. As these pools shrink, they

All sail set, an American wood stork (opposite) floats through a sea of air on wings flashing a bold black and white across five feet of Florida sky. North America's only stork, the "wood ibis" can flap along at nearly 35 miles an hour, rise on updrafts to dive playfully and rise again — or startle observers by rolling over to glide briefly upside down.

Thermals over Bharatpur, India, buoy a squadron of painted storks skyward (below). At the "effective top" — perhaps half a mile up — where the pull of gravity matches the push of rising air, the birds set their wings to glide effortlessly to distant feeding grounds or to another escalator of air.

BELOW: M. PHILIP KAHL
OPPOSITE: FREDERICK KENT TRUSLOW

concentrate the fish more and more. The storks respond to this early winter change in food availability. I was surprised to find that they usually begin nesting in December, when most birds are months away from their breeding season.

During my first season in East Africa, I was puzzled to learn that the yellowbilled stork starts breeding when the rains flood its feeding marshes, rather than during the dry season. I had expected the yellowbill to conform to the pattern of the wood stork, because of their very similar feeding behavior.

But as I became more familiar with the ecology of the area, the apparent paradox was resolved. The yellowbilled stork colony I studied was located near the northeastern shore of Lake Victoria, the largest lake in Africa. During a normal dry season, the marshes along the lake margin, where the storks fed, dried up completely. The fish were forced to retreat to the deeper waters of the lake itself —where, of course, they were unavailable to the birds. Only when the rains flooded the lakeside marshes did the fish once more enter the shallows. So here, as in Florida, the breeding season was timed to coincide with the highest annual availability of food.

Tricks of the trade help keep stork bellies full. Easy rider on a heedless hippo, an African openbill (above) snatches snails turned up by its mount in a wallow in Uganda. Like their Asian counterparts (right, upper) they wield their bills as forceps, often shucking a mussel or snail without breaking the shell.

Fishing by feel (center), a yellowbill wades Lake Nakuru in Kenya. A foot stirs cloudy water to scare up unseen prey. Now and then a wing fans out to "herd" fish to the groping bill. Lightning reflexes snap the mandibles shut on anything that scurries between them.

The marabou of Africa (lower) steals to survive. But not until the vultures tear open a carcass can the marabou snatch a meal for itself.

M. PHILIP KAHL

263

Africa is well known as "the land of tooth and claw." Alongside the immense herds of grazing mammals live the predators—lions, leopards, cheetahs, hyenas. As in any area where large numbers of animals are killed for food, abundant scraps are left over for the scavengers. The marabou stork has adopted the ways of the vultures, often dining with them at carcasses left by carnivores. Marabous are also ever-present guests around slaughterhouses and fish camps, where they glean a living from the refuse.

The marabou is obviously a relative newcomer to the scavenging game. Though the marabou towers over the vultures, it is less adept. The vultures' strongly hooked bills are adapted for tearing flesh. Lacking such a bill, the stork must stand by and wait for the birds with the proper utensils to cut up the meat.

In southern Kenya I watched waiting marabous suddenly dance in and steal morsels from the fighting, hissing, milling mass of vultures that covered a wildebeest carcass. Such thefts are no easy task, for the mild-mannered marabou is often intimidated and driven away by the more assertive vultures. Although the marabou could easily dispatch a vulture with one blow of its massive bill (they have reportedly killed African children this way), I have never seen one try. More aggressive behavior might gain an individual marabou fuller meals, but it would spell a dark future for generations yet to come. For where would the marabou be without the vulture to cut up his food for him?

I once kept a captive adult marabou that thrived for five months on nothing but scraps of meat. But the rapidly growing young need a diet high in bone-building calcium during their three months in the nest. Thus, breeding marabous alter their feeding habits, and approximately 30 percent of the food they procure for the nestlings is composed of fish, frogs, mice, and other whole vertebrate animals whose bones supply the needed calcium.

To collect this fare, marabous commute from nesting colony to rivers, marshes, and ponds shrunken by the dry season, when marabous normally nest. Then, as in Florida, isolated ponds teem with trapped aquatic animals whose density goes up as their habitat shrinks. Dry weather also means grass and brush fires; attracted by the smoke, marabous and other birds patrol the fire's leading edge and catch small animals flushed or injured by the flames.

White storks watch for an easy meal as an African grassland goes up in flames. Insects and small animals flushed by the blaze fall victim to these migrants from Europe—and to many other kinds of birds as well. Smallest of the storks, the two-foot-tall Abdim's stork shares the fire-chasing habit; so does the marabou, a giant of the stork family nearly five feet tall.

ALAN ROOT

Like most meat-eaters, all storks sometimes face food shortages. When this occurs during the breeding season, the nestlings may perish if the parents do not regurgitate onto the nest floor enough food to go around. The storks, like other predatory birds, have evolved a partial solution to this problem. Because incubation starts before all the eggs are laid, the eggs hatch at intervals of one or two days. Thus, the hatchlings appear in a "stairstep" fashion, the oldest as much as a week older (and larger) than the youngest. If a food shortage occurs, the larger young gobble it all and the smaller, weaker young starve almost at once, to be flung from the nest or even eaten by an indifferent parent.

A cruel and heartless mechanism to human eyes, perhaps — but it works, for it assures that at least a few offspring will survive under all but the most rigorous conditions. If all hatched at once and were of equal size and strength, they might divide the food equally and all starve together. The Australian zoologist A. J. Nicholson calls this the principle of "contest," where one wins and one loses, as opposed to "scramble," where both may lose.

Smallest of storks, the Abdim's stork is an inveterate insect-eater. As it follows the rains back and forth across tropical Africa with the seasons, it is seen by the local people as a harbinger of rain. Actually, it is following the abundant "flush" of insects brought out by the rains after a prolonged drought.

Especially in its breeding range — across sub-Saharan Africa from Senegal to Ethiopia — the return of the Abdim's stork in April or May is heralded by the natives as an omen of a good growing season. The more abundant the storks, so the belief goes, the more copious will be the rains. In these areas, the Abdim's stork fills nearly the same role in local folklore — a "superstition niche," we might call it — as does the white stork in Europe. Each is welcomed as a sign of spring and of good luck in general. And, as if to complete the analogy, the Abdim's builds its nests on rooftops in rural villages, as does the white stork in Europe.

All storks are masters of the art of soaring. Even in the tropics, where the stork family probably originated, soaring is essential if the birds are to forage far afield each day and then return to the nesting colony or roost at night. Birds such as the yellowbill often must fly as much as 60 miles round-trip on each expedition to reach the high concentrations of fish their feeding methods require. A seven-pound bird flapping this far would use up much of the energy gained in feeding. Instead, these birds have evolved the ability to tap the free energy in columns of warm air rising from land areas heated by the sun.

In midmorning, yellowbills flap a few hundred feet into the air, then ride to several thousand feet on these thermals. From this height they can glide for many miles before they must ascend again on another thermal. Thus they cover long distances in vertical zig-zags with little effort, returning in the afternoon in the same way. A second shift may soar out in midafternoon, feed all night, and return on the first thermals of morning. Waiting for a thermal to form takes time, and so shortens their feeding day as compared with birds such as herons and ibises that flap directly to and from their meals. Apparently the saving in energy—the getting of new food costs them little—makes it worthwhile.

Most storks live in sunny climes where thermals are abundant on most days. During rare periods of solid overcast or cool windy weather, though, the birds must flap laboriously to their food sources. And they don't look happy doing it!

Thanks to their ability to hitch free rides on thermals, Abdim's storks can migrate back and forth across the Equator with the seasons, exploiting shifting areas of food abundance. The white storks and black storks of Eurasia cover more ambitious distances: a white stork that nests in Poland may winter in South Africa. The white and the black storks thus have avoided competition with their tropical relatives, not by radiating into unique food niches but by leaving the arena.

Their soaring skill has preadapted them for such migrations. Preadaptation is a common phenomenon in evolution. It occurs when a structure or behavior pattern develops as a solution to one life-problem and then—strictly by chance—happens to solve another. Instead of developing specialized feeding techniques, the white and the black storks have used their mastery of the air to evolve migrations that remove them from the tropics during the breeding months, when food demands are greatest.

A few pairs of white storks have recently taken up breeding quarters at the extreme southern tip of the African continent. There they nest during the local spring and summer (September through January) in a climate not unlike that of Europe. Removed from the tropics, these too avoid competition with other members of their family, just as their brothers do in Eurasia.

But this may also be a biological reaction to environmental changes in Europe. Many of the wetlands where the birds formerly fed have been drained or otherwise destroyed, and such modern-day obstructions as high-tension wires take their toll of flying storks that collide with them. Storks have disappeared as breeding birds in Switzerland (where they are being reintroduced), Belgium, and

Stormy courtship and rigid ritual forge a nuptial bond between white storks on a barn in Poland. Flapping and biting (upper), the male at first asserts his ownership of the nest, though sometimes pausing to copulate before driving the female away. Accepting her at last, he joins his new mate in a neck-doubling greeting (center), act one in a ceremony of bows, pirouettes, and bill clattering that occurs every time a mate returns to the nest. With another rattle of his bill he mounts her (lower). Except for hissing and machine-gun bill clacking, the white stork is mute; others, such as the marabou, have loud voices.

Some nests—twiggy masses as much as six feet in diameter—last for decades and even pass down the family line. (One of the pair in the photographs may have been hatched in this nest.) Yet the nest-building ritual prevails; to an already ample nest the female adds sticks fetched by her mate.
M. PHILIP KAHL

southern Sweden. And their numbers are dangerously low in Denmark, France, Holland, and parts of West Germany.

The annual departure of the white storks from Europe attracts little attention; the birds generally leave singly and silently. Most young leave before their parents, apparently finding their way to the winter quarters by some inborn sense of navigation.

As the birds travel toward the African wintering grounds, flocks form from the converging paths of many migrants. Since open water provides few thermals, the migrants skirt the Mediterranean. By the time they reach the Bosporus or Gibraltar, often thousands of birds are moving in great flocks. As many as 51,398 birds have been counted crossing the Bosporus in one day—with up to 11,000 of them in a single flock.

Moving mainly during midday when the thermals are strong, the birds progress slowly and take many weeks to complete the journey, which for some may cover more than 6,000 miles. Early mornings and late afternoons are spent feeding around the night roost. If food is abundant or the weather unsuitable for soaring, individuals may spend several days at each stopover.

The long trek is not without its dangers. Each year thousands perish along the way from starvation, accidents, hunters' guns, and inclement weather. Hundreds have been killed by a single hailstorm. And no one knows how many may succumb to the rigors of crossing the Sahara and the Sinai Peninsula. White storks that survive the deserts fan out in all directions, some pressing southward, others remaining in equatorial regions.

Thousands of birds—banded with numbered rings in their European nests—have later been found dead along their migration routes and in the winter range. From these recoveries, much of the information about the white storks' migrations has been gleaned. Some birds return to Europe carrying African arrows or spearheads embedded in their flesh. When we trace these missiles to the tribes that made them, we gain further insights.

Although the dangers are great and the mortality high, the migration habit persists. Powerful biological advantages must repay the birds that move out of the crowded tropics to raise their families—and to bring good luck, good rains, and bouncing babes to the house with a stork nest perched on its rooftop.

Watch on a Rhine tributary keeps a white stork at its lofty nursery in Ketsch, Germany. In the lore of Europe the white stork brings luck; when luck meant large families, the stork began bringing babies as well. Now luck eludes this bird of legend. Though towns raise nesting poles and even rear nestlings when tragedy claims the parents, the stork's numbers dwindle in Europe.
JOHN W. TAYLOR

269

Stephen T. Emlen

EXPLORING THE MYSTERIES OF MIGRATION

The honking of geese as they wing overhead in V formation...
the cacophony of ducks rising by the hundreds from a marsh at
dusk...the leap of a salmon in a foaming cataract...these are
sounds and sights of migration, a spectacle that has intrigued men
for thousands of years. Yet only in the past few decades have scientists begun answering such questions as where do different animals
go — and why. What tells animals when to migrate? What guides
them along their pathways?

Impressive migrations occur throughout the animal kingdom.
European eels hatch in the Sargasso Sea and cross the Atlantic to
mature in freshwater streams of Europe; adults are thought to return
to the Sargasso Sea to spawn. Salmon in the Atlantic and the Pacific
complete the opposite cycle, hatching in fresh water and maturing

Summoned by the circling
seasons, migrant black skimmers
wing sunward on an ageless
quest. What beckons earth's
many migrants, what guides them
afar, what calls them back?
In this composite photograph
a modern marvel seems to sift
for clues as the skimmers pass.

Between space shots, NASA's
radarmen at Wallops Island,
Virginia, have locked this
five-million-watt radar onto
a single migrating bird released
by the author. The 60-foot dish
antenna can track the bird for
40 miles, gauge speed, altitude,
course — and even wingbeats.

Yet mystery remains, as myriad
creatures throughout the world
heed the primal urge to move on.

BLACK SKIMMERS IN THE EVERGLADES
BY OTIS IMBODEN, NATIONAL GEOGRAPHIC PHOTOGRAPHER;
WALLOPS ISLAND RADAR BY NASA

at sea; their return to spawning beds is remarkably precise, for it may take several years and cover several thousand miles.

Most baleen whale species leave polar waters to calve in more temperate climes, then return with their young to seek out rich pastures of krill. From faraway feeding grounds green turtles return time and again to the same beaches to drag themselves ashore and lay eggs. The oceans are full of such experienced travelers.

Land animals also migrate. Some bats leave breeding caves of the temperate zone in autumn, travel to warmer areas, and return in spring. Winter on the tundra drives herds of North American caribou to areas where snow is shallow and easy to paw through for browse; some deer and antelope also journey afar. Even insects migrate; many dragonflies and butterflies make impressive trips.

But the best-known migrants are the birds. Fully two-thirds of the species summering in the northern United States and Canada travel to the southern states and Mexico, or even to Central and South America, for the winter. Similar movements toward more tropical areas take place throughout the Northern Hemisphere and to a lesser extent in the Southern Hemisphere. Many birds, particularly shorebirds, fly as far as 12,000 miles to their winter homes and back.

Most bird migrants tend to nest in the same areas year after year — and to return faithfully to the same wintering spots. Many of them make the journey alone and at night. Details of such behavior are fascinating. So is the fact that it happens at all.

A migration trip, whether of 600 or 6,000 miles, requires considerable physiological preparation. The high cost in energy, the risk of being buffeted about or blown off course, the problems of finding food and avoiding predators while traveling through a wide spectrum of habitats—all argue against the likelihood that the phenomenon of migration would ever arise. Why then do birds migrate?

Some scientists suggest that Ice Age glaciers first forced birds from ancestral nesting areas, to which their descendants began to return when the ice started to recede about 20,000 years ago. But there is no reason to believe that migrations were not taking place prior to the last glacial advance. And most long-distance bird migrants belong to groups that probably evolved not in temperate latitudes but in the tropics, well beyond the reach of the glaciers.

Another theory holds that birds simply retreat each autumn from the cold and the dwindling food supply. But many birds leave their breeding areas long before bad weather sets in or food declines. Even in equatorial regions, where there are no such hardships, extensive migrations occur. So we must ask why birds leave the tropical areas at all. Why don't these (Continued on page 279)

The bug that caught a bird: A tiny radio rides a Swainson's thrush on a flight for science. Gentle hands fit the migrant— netted over upstate New York— with a transmitter weighing about a tenth of an ounce. Ballooned aloft in a trap-door box, the bird drops free at a preset altitude, usually 4,000 feet. Author Emlen then tracks the radio signals as the bird orients itself high above distractions; its course can be plotted for 15 to 25 miles.

Mission accomplished, the bird can peck at the radio's flimsy wire harness and free itself in a day or two; none held in captivity has ever failed to get the bug off its back.

To test whether some migrants steer by earth's magnetic field, scientists send up captives wearing tiny magnets that would disrupt such a guidance system. Radar can track the birds to see if they fly off course. Soon they shed their glued-on cargo to steer by senses whose secrets men probe but do not yet know.

ANTHONY A. BOCCACCIO

*March to motherhood draws a dotted line of pregnant caribou and
a sprinkling of yearlings on a snowy slope in Alaska's Brooks Range.
In spring they hurry ahead of the herd to calving grounds on bleak uplands
perhaps 300 miles from the winter range, then lead the calves
downslope to rejoin the herd for a brief summer's sojourn in lush lowlands.*

STEVEN C. WILSON

Migrants of sea and sky shuttle with the seasons

Blackpoll warblers congregate each fall along the New England-Canadian coast; some have already flown 3,500 miles from Alaska. Moving on, they traverse 2,100 miles of ocean to South America. Most don't even land at Bermuda, sole waystop in the North Atlantic, unless they are weathered in.

European eels of the genus *Anguilla* hatch in the Sargasso Sea and ride the Gulf Stream on a two-and-a-half year voyage to Europe. Their leaf-shaped bodies swell to cylindrical form as they enter fresh water to mature. Perhaps a decade later adults re-enter the ocean. Presumably they make their way back to the Sargasso Sea, spawn in its deeps—and die.

Golden plovers span both Atlantic and Pacific. Some birds of the Atlantic subspecies may fly a loop that can be 15,000 miles—from boreal beaches to Brazil, Uruguay, and Argentina, returning via Central America. Young birds fly the land route both ways; adults rarely rest on the water to ease crossings.

White storks from Europe funnel past the Bosporus and Gibraltar en route to Africa; few cross the Mediterranean. Awed observers at the Gulf of Suez may see thousands of storks flying low over the water in a narrow stream many miles long, as they make an autumnal migration that will take many of the birds as far south as South Africa.

Green turtles tagged on Ascension Island reappear feeding off Brazil or breeding later at Ascension. From Brazil the island, 1,400 miles away in mid-Atlantic, is a tiny target. A miss would take a turtle on to Africa—where no Ascension turtle has ever been observed.

Humpback whales in Arctic and Antarctic seas feed in herds on summer's plankton. Perhaps spurred by shortening days, the whales swim thousands of miles to breeding grounds off warmer coasts; females calve and mothers with yearlings wean them and mate again. Until the whales return to polar seas in spring, only sucklings eat regularly.

DRAWING BY NANCY SCHWEICKART, GEOGRAPHIC ART DIVISION

migrants avoid hardship by remaining permanently in overwintering areas? Why return to the temperate and polar regions?

Because relatively constant conditions prevail in the tropics, the number of birds there is close to the carrying capacity of the environment; competition is high for food and space. But competition is less in temperate areas, where fewer species reside and where seasonal changes allow a summer bloom of plants and insects. Thus a migrant that times its trip to take advantage of this seasonal food abundance should be more successful and rear more young than if it were to remain and breed in the crowded tropics.

But it would be unwise to overgeneralize. The migratory habit has evolved independently among many kinds of birds. Different species travel in different directions, to different places, at different times, and for different reasons. Whatever the causes, migration would not have evolved unless the benefits exceeded the hazards.

The Pacific subspecies of the golden plover breeds in the Alaskan tundra. Each fall these shorebirds travel 3,000 miles—probably nonstop—over the Pacific to Hawaii, then continue south for another 2,500 miles to the Marquesas, and often 500 miles more to the Tuamotu Archipelago. For most of the way over the featureless ocean there is no land on which to stop and rest.

Large birds like the plover are not the only ones that perform such feats of navigation and endurance. The little blackpoll warbler leaves the northern forests of Canada and Alaska to gather along the Atlantic Coast in New England and Canada, then heads overwater to South America. Even the tiny ruby-throated hummingbird —a creature scarcely three inches in length—routinely flies 600 miles across the Gulf of Mexico on its migration between the eastern United States and Central America.

Consider the energy problems of such flights. Physiologists believe a flying bird uses six to eight times as many calories as one at rest. How can a bird fly such great distances nonstop?

The fuel for these flights is fat. Just before migration the bird's metabolism changes, and large quantities of fat are stored in the body. Migrants often are bulging with fat layers; a blackpoll netted just before it begins its overwater flight may weigh 3/4 of an ounce, of which fully half is stored fat.

Birds conserve energy by taking advantage of the winds. There is growing evidence that birds are excellent meteorologists, timing departures to catch favorable winds—behind a cold front in autumn, for example, or ahead of a warm front in spring. This guarantees that they will rarely be aloft in strong headwinds or crosswinds. But an incorrect forecast can spell disaster. Bad weather often grounds

Traffic jam roils the Adams River in British Columbia (opposite) as sockeye salmon in nuptial hues struggle upstream on a life and death journey—life for eggs they will leave behind, death for themselves when mating ends. Like other salmon in both Atlantic and Pacific, their spawning run ends in the same stream, perhaps the same riffle, where life began years before. Thence a young sockeye works its way to sea and loops in year-long circuits until driven home by one of nature's strongest instincts. Somehow the salmon finds its river, and with nostrils that can detect odors diluted one part to a billion it seeks the stream whose unique scent it may recall from infancy. Dams, pollution, bears, and men stop many. One that got away leaps up a cataract in Alaska's Brooks River (below) to its fate.

ROBERT F. SISSON, NATIONAL GEOGRAPHIC PHOTOGRAPHER

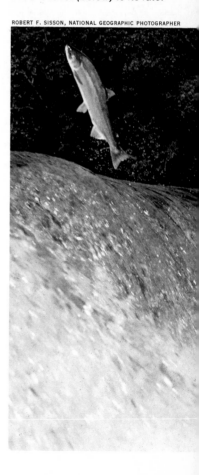

*Migrant monarchs bloom in black and orange on
a cypress at Stinson Beach, California.
Though 100 of the butterflies weigh but an ounce,
thousands bend branches as they rest in
their winter refuge. Each fall monarchs soar from
northern states and Canada to winter as far
south as Mexico. Year after year new generations
find the same locales, even the same trees —
without ever having seen them before.*

*Tags help map their travels. One bore its tag
some 2,000 miles from Canada to Mexico. Others
may span seas on ships or airplanes; monarchs
have appeared on Pacific islands and in England.*

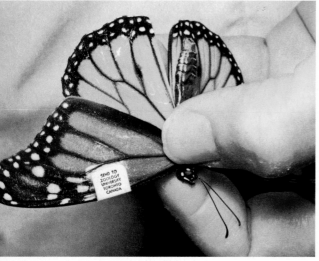

birds in environments where they are inept at finding food or avoiding predators. A sudden night snowstorm in Minnesota once overtook Lapland longspurs, small birds that migrate to and from polar regions. The next morning an estimated 750,000 dead longspurs littered an area of one square mile.

We can only guess at the fate of birds that encounter harsh conditions over open water. A few alight exhausted on ships far out at sea. If the autumn weather changes suddenly over the Gulf of Mexico, birds aloft may have to buck heavy headwinds to reach the beaches of Yucatán. Observers there have found some so exhausted they could be picked up by hand. The birds were completely devoid of fat; some had even used up much of their own flight muscles as a last possible source of fuel.

The disastrous effects that bad weather can wreak help explain why birds generally are so attuned to meteorological cues. But such adaptations are of no use if the birds cannot find their way between breeding and wintering areas. How they are able to do it is a question that has inspired scientific investigations throughout the world.

Every good field ornithologist knows where he can find large numbers of birds during migration seasons. Usually he looks for some feature of the topography that draws concentrations of birds. Daytime travelers, for example, frequently follow river valleys or mountain chains. They show a great reluctance to cross a large body of water and will turn instead to follow the coastline. Such tendencies result in masses of birds at such famous bird-watching localities as Point Pelee on the Ontario shore of Lake Erie and Cape May at the southern tip of New Jersey.

But after dark the importance of the terrain diminishes. Radar studies have revealed that nocturnal migrants generally ignore even the most prominent landscape features beneath them. Then what cues are available to the night fliers?

Gustav Kramer, working in Germany in the early 1950's, made the first important breakthrough in the study of bird orientation. His ingenious experiments showed that a caged starling would flutter in the direction of its normal migratory route—as long as the sun was visible. When the sun's position was deflected by mirrors,

ANTHONY A. BOCCACCIO. OPPOSITE: STEVEN C. WILSON

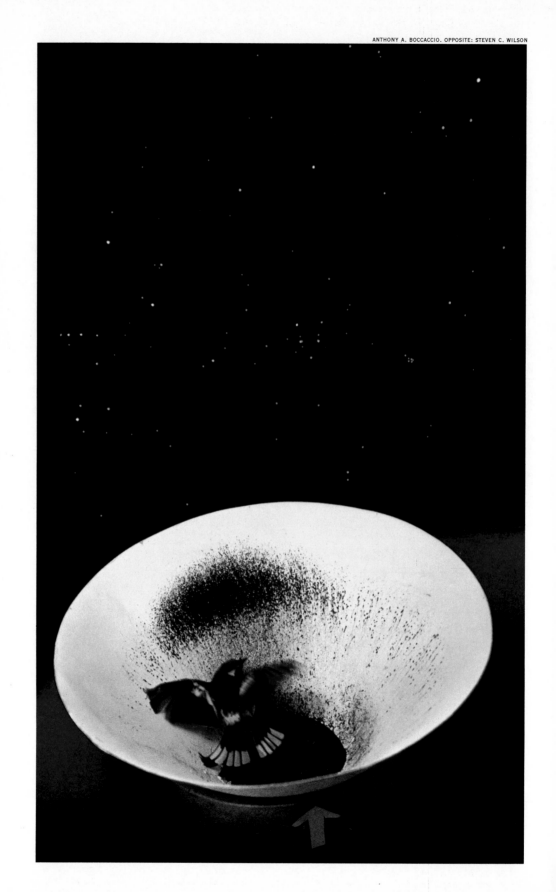

the bird changed its direction and maintained the same angle relative to the sun's image. And when the sun was obscured, the starling became disoriented and fluttered in random directions.

Later experiments demonstrated that birds make use of an internal biological clock that enables them to compensate for the apparent movement of the sun across the daytime sky. This integrated time sense is essential for maintaining a constant heading throughout a long flight. The ability to use the sun as a compass appears to be widespread in the animal kingdom, not only among birds but also among insects, fishes, and amphibians — and probably sea turtles. But we still have not determined how an animal knows in which direction it must go.

Studies of caged birds at migration times suggest that nighttime travelers steer by the stars. On a starry night they concentrate their fluttering in the correct migratory direction, even when their view is restricted to a portion of the sky. Under overcast skies, however, they usually flutter at random.

In principle there are two ways to orient by the stars. A bird might locate one star or group of stars and fly at a particular angle relative to it, adjusting this angle with the same kind of internal time sense that enables the daytime traveler to steer by the sun. Or it might use the constant patterns of stars to locate a particular direction — just as we sight along the pointers of the Big Dipper to find the North Star. As long as such a pattern is visible, we can determine north without knowing the time, the season, or our location. Different species of birds appear to employ one or the other of these

Ink on its feet, stars in its eyes, a magnolia warbler writes a page in the fascinating story of animal migration. Building on classic experiments and using Cornell's planetarium, the author placed various migrants in blotting-paper cones topped with transparent lids and floored with ink pads. Footprints on the paper showed each bird's orientation as it tried to fly.

Under real autumn stars most jumped southward; others, tested in spring, inked the north side of the cone. In the planetarium, when the spring sky was reversed, so were the birds; most jumped toward a man-made Polaris that actually beckoned them southward. And when men switched off the heavens, the birds hopped at random — strong evidence that some birds steer by the stars.

Others may not. On a cloudy night, sandhill cranes, en route from nesting sites in Alaska, point a perfect V toward Mexico as they print silhouettes on an Oregon sky. Radar has even tracked migrants flying on course between clouds that blotted out sky and ground.

methods. I have found that indigo buntings use the latter, concentrating on star patterns in the northern sky. European warblers seem to rely on integrating a time sense with the positions of selected stars.

Many persons point to such studies and conclude that the mysteries of animal navigation have now been solved. This is not true. Current theories fail to tell us how or why an animal selects one direction over another. And recent evidence indicates that star and sun cues may not even be essential for some migrants.

During the past two decades ornithologists have been observing migration by radar. A bird aloft is a tiny target but, if it is big enough and close enough to a tracking antenna, it can produce a blip on a radar screen. We can determine the density and directional tendencies of such blips and correlate them with the weather. These studies show that some migrants remain well oriented under total overcast and even between cloud layers. This does not disprove that birds use celestial information, but it certainly implies that they may also use other cues for directional guidance.

We do not know what these cues may be. Might a bird use the wind as a cue? Might some even orient by the earth's magnetic field? Recent studies suggest both of these possibilities.

Perhaps one day we shall know. Thus far, most of our studies have involved one of two techniques. We have conducted experiments in the laboratory, where flight conditions cannot be duplicated nor the full spectrum of normal behavior observed. Or we have watched migrants in the wild, where we could not manipulate them and where we could not demonstrate a cause and effect between orientational cues and the behavior we saw. But it is now becoming possible to combine these two techniques.

We can track a bird in flight by attaching a miniature radio to it. Better yet, an investigator can use powerful, narrow-beam tracking radars that lock on and automatically track an individual bird. Either way, a captured bird can be exposed to various manipulations, released under selected weather conditions, and then tracked as it actually makes its navigational decisions.

This new approach with free-flying migrants should help us toward a better understanding of the navigational systems of animals. We still have much to learn. Our knowledge is insufficient to explain why a blackpoll warbler flies out over the Atlantic to make a nonstop voyage to South America. Our studies have not proven how a marine green turtle leaves its feeding grounds off Brazil and finds, 1,400 miles away, that pinpoint of land known as Ascension Island. This we do know: The enigmas of animal migration will continue to fascinate mankind for generations to come.

Plowing a wake in the sand, a green turtle heads for the sea, leaving some 100 golf-ball-size eggs buried in this Costa Rican beach. Tests show she finds the sea by the brightness of its unbroken horizon; fitted with light-filtering goggles, turtles blunder off in zigzags even when a few feet from the surf.

Scientists seek to learn how adult green turtles that feed in Brazilian waters find tiny Ascension Island. In its waters they mate; on its beaches they lay their eggs. With eyes not sharp enough to steer by stars, they may orient by the sun and home in on the island's "odor," wafted westward by constant currents.

These same currents would carry the hatchlings from Ascension to Brazil in a few weeks—yet none younger than a year has been found there or anywhere else. Where little turtles go is but one of the many mysteries of migration.
ROBERT E. SCHROEDER

THE RITES OF SPRING

John Hurrell Crook

Spring, the sweet Spring, is the year's pleasant king;
Then blooms each thing, then maids dance in a ring,
Cold doth not sting, the pretty birds do sing—
Cuckoo, jug-jug, pu-we, to-witta-woo!

Spring indeed seemed to burst upon the Elizabethans of Old England with the clatter of birds in high places, of lovers and lasses singing madrigals and cavorting on the greens. No wonder, in those high-beamed halls, when the warming sun glimmered down the drafty galleries to chase winter's chill! Then did the heart stir and the mind reflect on the marvel of it all. Few poets have not tried a verse or two in glory of what Thomas Nashe called "the year's pleasant king"—the season when life renews itself, when birds nest and lambs are born in the fields, when warblers and other wanderers return, and when the woods fill with foliage and insects.

Every year I too rejoice when, on waking early, I hear once more the rich voice of a familiar thrush. All over the Northern Hemisphere at this time the avian orchestra tunes up. In my mind's eye I visualize the slow melting of North American snows, the creep of green over frozen Europe, over Siberian wastes. I imagine winged hordes coming out of Africa to nest again in the freshly temperate north—specks of life often making the trek in one great nonstop flight, the miracle of the annual breeding migration.

Like many a naturalist, I began my career with a boyhood fascination for the events of the English spring. Bicycle rides at dawn led me to the nightingale, the "light-wingèd Dryad of the trees." There were expeditions to Scottish islands to see great seabird colonies or to seek the golden eagle—sound training for later travels. Since then I have worked in India and Africa, studying in particular the gregarious weaverbirds and those brawling knights of Ethiopia's high Simen crags, the gelada baboons.

Every land has some kind of spring, however brief. Even seemingly unchanging tropical regions have periods of greater leaf renewal after a shedding. Nowhere, except perhaps in the deepest recesses of caverns or abysses of the sea, is there a truly seasonless ecology. Most of earth's surface is subject to major climatic shifts as the sun's heat swings north and south with the changing year. Nearly everywhere one can witness the importance of the seasons for animal life—timing migrations, regulating onset of reproduction, imposing shifts in social relations between individuals and in the organization of whole societies. Indeed, the sun's passage regulates the great symphony of life, drawing with it, as the moon does the tides, the urge for replenishment, for creation, for reproduction.

But why should mechanisms responsible for life's renewal gear themselves to the timing of spring? Why not summer or autumn? And how do those mechanisms work?

In courting grebe or hidden world of cell and chromosome, life is sustained by the urge to renew. 287
PAINTING BY RICHARD SCHLECHT

The answer to the first question is simple. Animals breed at those times of year that offer the most promising conditions for raising young. In particular, animals tend to breed when food resources not only are adequate for the adult population but also exist in surplus quantities. If not compelled to forage constantly, animals can have the spare time that courtship demands. And plentiful food provides the energy necessary for reactivating dormant sex glands, for egg manufacture, for incubation and feeding young in the nest—or, in mammals, for pregnancy and lactation. Not all newborn need attentive parents, yet the principle remains the same: Fish, frogs, and reptiles produce fertilized eggs at a time when developing young can find enough food to grow to maturity.

Spring in North America and Europe is a remarkable period. As the foliage opens, the insects appear. Later the flowers seed, the grassheads grow and ripen, the land seems glutted with food resources. Yet these often do not endure for long; the period of abundance ebbs, and lean times unsuitable for rearing young begin again. Each animal species has its own characteristic way of exploiting food. There are insect specialists, grass-eaters, fruit-eaters, meat-eaters, and there are some, omnivores, that eat foods of all kinds. The

timing of each species' breeding season is related to peak abundance of its type of food. In England, for example, wood pigeons return to reproductive condition gradually, with most birds not breeding until late summer, when the cereal grains on which they feed their young begin to be plentiful. Their close relatives the stock doves, by contrast, take seeds of weeds that mature quickly and keep proliferating; they can rear young earlier, and thus have a longer breeding season, than wood pigeons.

Studies suggest that birds make use of environmental signs—warming temperatures or the onset of green after rain, for example—that herald the coming of good feeding conditions. Indeed, birds that fail to respond will be unable to raise many, or any, offspring. Their stock soon must be eliminated. So, for each species, natural selection ensures the evolution of a sensitivity to the environment—a sensitivity setting off physiological mechanisms that prepare the species for breeding at the best time.

These mechanisms have their seat in an area of the forepart of the brain called the hypothalamus. Here the brain's responsiveness to external factors—such as the lengthening days of spring—triggers the activity of that master gland, the pituitary. It sends chemical

Beaver-busy, a village weaver adds a green wisp at the entrance to his nest. Males have a compulsion to build, plaiting one nest after another to lure females. The polygamous males then rip unused nests apart and start over. A bird begins a nest frame in a palm tree with strips torn from a leaf by seizing a beak-hold and flying away. Weavers can skeletonize a tree, hanging scores of nests at frond tips. Colonies prefer sites near man's abodes, where predators such as hawks and snakes are scarcer.
  More than 100 species of weavers spread from Africa to Malaysia. Some build a vast multi-nest structure whose common roof all keep in repair. 289
EDWARD S. ROSS

messengers coursing through the bloodstream. One of these, known as luteinizing hormone, or LH, awakens the winter-dormant sex glands. Another, called follicle-stimulating hormone, or FSH, brings about the formation of sperm and eggs. The precise functioning of this apparatus varies from species to species.

One experiment studied the effect of different light schedules on the sex organs of male greenfinches. An artificial "day" of 7 hours of light and 17 of darkness was compared with such a cycle as one of 6 hours light, 7 dark, 1 light, 10 dark. Results indicate that short periods of daylight are adequate to produce the LH responsible for revitalizing testicular tissues. But additional hours of light are required for the FSH that is needed for effective production of sperm. Thus duration of the light periods is a prime factor in the timing mechanism of avian breeding.

Following breeding, sex glands cease to respond to environmental stimuli. They become "refractory" and no reproduction is possible so long as this state continues. Evidence suggests that the refractory period in birds is commonly initiated by the bird's response to summer day-lengths. While the increasing day-lengths of spring bring about reproductive condition, long summer days turn it off again.

Food supplies have had a role in the evolution of the refractory period, which varies among species. The chaffinch of Europe, for example, breeds in April and May. It rears

Shaped by a mighty solar engine—and plotted on this abacus-like device—the patterns of breeding behavior among birds reveal how sunshine and sex intertwine. Earth's slanted axis tilts hemispheres alternately toward the sun, bringing the seasons. Spring ushers in longer days that become a factor in waking dormant sex glands. Mating and brooding follow, timed in each species to coincide with seasonal food abundance. On the abacus, the rods represent broad groups of north temperate zone birds. Rod C depicts species in which male and female stir

TASI GELBERG PESANELLI, INC.

a single brood, then enters a long "barren" spell—months when insects it feeds its young are not plentiful. But the greenfinch, which brings fledglings summer's seeds, has a long breeding season and a short refractory period. And it can rear more than one brood.

Both the chaffinch and the greenfinch cease to be refractory in late autumn. In fact, the day-lengths of fall, matching those of spring, are commonly sufficient to start the renewal of breeding behavior by most birds. But now other factors intervene to prevent reproduction. These inhibitors include steadily dwindling hours of daylight, dropping temperatures, and poor food supplies. And so avian sex glands slip into winter's resting state—to await another spring's resurgence.

The timing of breeding seasons is not always a response to changes in day-lengths. In equatorial areas, where the daily ration of light changes little with the seasons, the onset of the rains serves as a stimulus. Once in northern Uganda I witnessed the passage of local storm clouds over the vast landscape. As I drove into an area within the rain path of one of these clouds, I found a host of weaverbirds busily setting up nests in trees over freshly formed pools. They appeared to be flying in from the surrounding dry country expressly for that purpose. A few days later I passed by again. The water had evaporated, the nests were drying, and most of the birds had gone. Only when several succeeding rainfalls set vegetation and insect life flourishing does breeding in such areas reach full

sexually as daylight begins to lengthen, pair to rear one brood in the spring, then enter a long "refractory" period— a safety-valve stage barring unseasonal hatchings. In England the rook, for example, breeds when earthworms are plentiful and ends before hot weather drives worms too deep to catch. Rod B birds need long days to trigger mating; they hatch a midsummer brood. Birds of Rod A type, less bound to one food peak, can breed more than once. Those of Rod D, with an even more varied diet, may breed during a span as long as ten months.

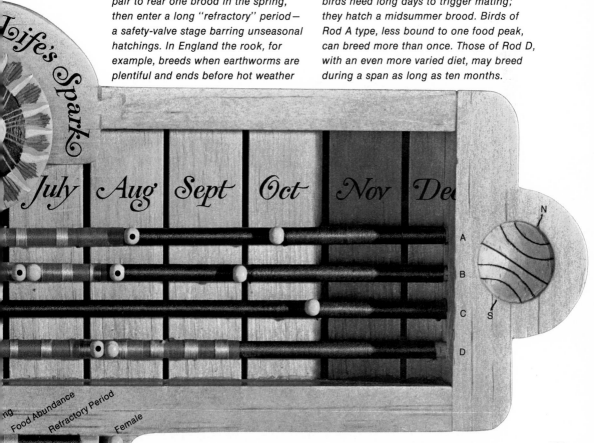

Life's Spark

July | Aug | Sept | Oct | Nov | Dec

A
B
C
D

N
S

Food Abundance   Refractory Period   Female

291

Bloodied male elephant seals duel for dominance. Pendulous noses are part of a resonating system that blares their snorts for miles. Such frays—and far commoner bloodless threat displays—set a breeding site "peck order." Dominant males monopolize access to cows; a tough beachmaster may keep all other males from mating. The fittest thus enhance the species' genetic heritage and, in polygamous animals, may shape male-female contrasts—sexual dimorphism. Two-ton bull elephant seals are twice as big as cows.

scale. For some dry-country birds, breeding may be completely suppressed in rainless years.

Rainy seasons also trigger reproductive activity in other animals. Frogs often seek pools formed after a drought. Reports of frog "wars" can be traced to breeding frenzies brought on when the aestivating animals rise from the mud following a heavy rain. In one such incident near Kuala Lumpur in Malaysia an estimated 10,000 frogs and toads of ten species roiled a shallow pool some 15 feet in diameter.

Reproductive behavior goes on in a rhythmic cycle even in those environments where a year's seasons appear almost uniform. On Ascension Island in the Atlantic Ocean near the Equator, sooty terns breed at intervals of nine and a half months. Brown boobies breed every eight. Other seabirds in other tropic regions follow similar cycles. Why, with light and seasonal changes at a minimum, should there be such a rhythm? Why, in fact, should the mating rites of individuals be synchronized with others of the species at all?

Where no environmental factor favors reproductive success at a particular season, species able to breed with greatest frequency have an advantage. Such species tend to go through the full breeding sequence as quickly as possible. Their rhythm is internal and does not coincide with the cycle of the year. Social stimulation in the colony probably prompts all members to activate their reproductive organs at about the same time. And when the breeding time is the same for all birds in a colony, one theory holds, the presence of so many young increases the odds against an individual being taken by a predator, and the predator's inroads won't be enough to affect the species' survival.

Stimulation between individuals does dramatically affect the physiological preparation for mating and parental behavior. Presence of a male canary, for instance, is vital in bringing about his mate's nest-building. Male ring doves develop milk-secreting crops partly as a consequence of watching their mates brooding. This intricate relationship between behavior and hormones has been studied in varied experiments. Stimuli from the male's courtship — itself triggered by hormones — are crucial to the development of the female's sex glands; female ring doves caged with a view of courting males produce more female hormone than those that can see only inactive, castrated males.

We have been looking at the breeding seasons of birds simply because so much is known about them. Even in the oceans, though, marine invertebrates respond to seasonal rhythms. For these animals, a rise in the water temperature becomes a stimulus often more important than light. Flushes of new life occur at later dates as one moves north along the coasts, or from the shallows to colder, deeper waters offshore.

Occasionally the breeding rhythms can be remarkably precise. On coral reefs off Japan, off the Dry Tortugas in the Caribbean, and off Samoa in the Pacific lives the amazing palolo worm. By day it hides most of its 16-inch body in tunnels and crevices, reaching out with its head at night in a search for food. But once a year, with a timing geared uncannily to phases of the moon, the palolo backs out of its burrow. A vigorous twist breaks away the hind portion, which wiggles tailfirst toward the surface. The shorter head-end remains — over ensuing months to regenerate another tail.

Swimming rapidly, the tail sections swarm at the surface as thickly as noodles in soup. Those from male worms are reddish-brown or yellow, those from females a bluish-green or gray. Then, their one- to three-hour swim ending with dawn's light, the tails burst open.

293

Body walls disintegrate; eggs and sperm mingle in a fecund broth. In time larvae sink to the bottom—and a palolo cycle starts again.

Strange cycles occur on land too. Studies with stags of the red deer on the Scottish isle of Rhum reveal that in spring and summer, when antlers are in velvety growth, no male hormone is detectable in the animal's blood. When the hormone concentration rises, antlers harden; stags joust head-to-head, antler size establishing the order of dominance. Then in mid-September the hormone level jumps abruptly. The stag's voice becomes a roar, his neck thickens; he thrusts with his horns at the heather, roams in search of a mate. Now body size and neck strength count for more than antler spread in ritualized fighting between males. In time the hormone dwindles and antlers drop off. Battles become boxing matches; the stags spar on hind legs, and the social order shuffles anew.

A mong mammals, females shelter the embryo within their bodies for extended periods. The timing of birth involves commitment to an intricate series of physiological processes which begins as soon as the embryo attaches itself to the walls of the uterus. The goal is to produce the young at the most favorable time. In certain species of seals, such as the Stellar sea lion and the northern fur seal, this has resulted in a peculiar adaptation. These animals roam the oceans for most of the year, then congregate in early summer. Mating occurs soon after the cows have "hauled out" from their year's wanderings and given birth. This is convenient, since otherwise the males and females would have to seek each other over the trackless sea. But only eight months are needed for development of the embryo, and this would mean that births would occur at an unsuitable time. Seal cows "solve" the problem by carrying the fertilized egg within their bodies in a kind of suspended animation. Attachment of the egg to the uterus—implantation—is delayed until eight months before the ideal birth time.

The naturalist's fascination with such exquisite mechanisms of the rites of spring leads to questions at biology's heart: Why should there be two sexes at all? Are there other kinds of reproduction? How do sex roles affect animal societies?

Simple living things like bacteria and viruses reproduce by dividing in half. Yet sexuality appears even among single-celled organisms—the majority, in fact—and it involves a mixture of genetic elements from two sources. The advantage of such a process is that it allows new characteristics, arising through the mutation or recombination of hereditary materials, to diffuse gradually through the whole population. As conditions change on the surface of our planet, organisms will be successful if they develop reproductive systems allowing adaptation to change; other organisms will die out.

In the great majority of higher animals and plants, sexually produced individuals have two sets of genes, present in pairs. If the members of a gene pair differ, one is generally dominant—its characteristics shape the development of the individual. The other, "recessive," gene remains latent, wielding its influence only when teamed with a similar recessive gene or when it emerges through a chance mutation or some other genetic alteration. Now, if the environment changes, some of the genes from this recessive store may prove useful to the organism and ensure survival. Natural selection then brings a rapid increase in the proportion of such genes and the population adapts more quickly to the new conditions than would otherwise be the case.

 (Continued on page 302)

A NEED FOR A NEST

In crimson glory a frigatebird displays his throat sac for a mate he has enticed to a twig nest atop a shrub in the Galapagos Islands. Normally the sac is a shriveled strip of pink skin. But at mating time it turns scarlet; the male forces in air to make it a taut balloon.

The remarkable courtship mechanisms of the avian world are shaped by hormonal changes. And the changes themselves are triggered by responses to external stimuli. In many birds contact with nest materials prompts development of brood patches — breast areas rich in blood vessels — whose warmth aids incubation. Stimulation from nest and egg also prods mechanisms that affect laying. Removing eggs from an "indeterminate" hen's nest will cause her to keep producing beyond the normal number in her clutch; adding will make her stop early. Similar steps have no effect on "determinate" hens, which usually produce only the clutch's fixed number of ova. To rear young, a pair may stay bonded or either may toil alone.

For a bower, something blue

The bowerbirds of Australia and New Guinea build arbors as court-ing grounds and decorate them with anything from beetle wings to pilfered car keys. Among them the satin bowerbird rates as a master architect. The pigeon-size male erects twin rows of twigs—oriented almost invariably north and south—to form an avenue bower. Before it he displays objects he has collected or stolen from neighbors; he has a penchant for items colored blue. He paints the arbor's inner walls with plant juices or charcoal mixed with saliva; a plug of chewed bark keeps beak agape so the mixture can ooze onto the twigs. The male may show off one of his prizes to the greenish female he coaxes to the bower, then spread his wings in ritual dance. After mating, he drives her away (below). She will raise the brood by herself.

NORMAN CHAFFER

Dawn to dusk, a slave to a mound

Incessant toil under a baking sun burdens the life of the mallee fowl. Named for the scrub of its Australian habitat, it belongs to the Megapodiidae — birds that bury eggs for hatching. The male mallee's incubator is a mound as much as 15 feet across and 3 feet high. In a vegetation-packed chamber within, the female lays as many as 33 eggs at irregular intervals. To the male falls the daily chore of opening and closing the mound, temperature-testing with bill or tongue. Kicking away or adding sand regulates the effect of heat from the decaying vegetation and from the sun; the bird's diligence keeps chamber temperatures amazingly constant. Incubation takes about 50 days. Once out of the egg, a chick struggles to the surface and totters off, ignored by the adults. In 24 hours it can fly.

298

GRAHAM PIZZEY

Baby-sitting in a walled-up nest

Trip after trip the slim bird flies from water hole to nearby hollow tree, carrying pellets of mud in her enormous beak. At the tree sideways thrusts of her bill plaster the mud around a hole in the trunk, gradually narrowing the opening. Her somewhat larger mate flaps about solicitously. He frequently brings her fat insects, which she pauses to accept. At last a gap hardly more than an inch wide remains in the mud plug, and through it the female squeezes. Using mud that has dropped inside, she reduces the gap to a beak-size slit. Thus begins the imprisonment of the red-billed hornbill, one of the family of birds that rear their young in sealed chambers.

The walled-in life presumably provides protection against such predators as snakes and genet cats. The female hornbill carefully blocks all other crevices in the tree, compacting the mud with rapid swipes of her beak. Inside her nest she lays two to five eggs at intervals of about two days. While she incubates the eggs, she molts, pushing shed feathers through the entrance slit. Out the same way go her droppings, ejected in high-velocity squirts.

The male, believed to mate for life, feeds his imprisoned family; his 30-odd trips a day increase to more than 70 after the young hatch. The diet: seeds, berries, lizards, and insects; a green mantis makes a mouthful almost as big as one naked 6-day-old chick. Quarters get crowded when the pinfeathered youngsters are about 18 days old, and the mother may resort to stretching room in the hollow tree. Some three weeks after the eggs hatch—nearly six after sealing herself in—she pecks a hole in the plaster and leaves the nest. Immediately the young work to reseal the opening, using their own droppings and slugs and sticky berries the father supplies.

Beaks of the youngsters are a pale orange. Not until adulthood will they become the hornbill's rich red. Upturned tails ape the mother's posture forced by confines of the nest, but young chicks stay that way even when taken from the chamber. At about six weeks the hornbills are fledged. In turn they chip their way out and take flight, wings and legs flailing. These pictures were made in Kenya's Tsavo National Park through a pane of glass set in the nest tree.

JOAN AND ALAN ROOT

301

Many primitive aquatic animals reproduce merely by broadcasting into the water the sex cells which carry the genes. If male and female seek each other out and discharge their sperm and eggs simultaneously, chances of fertilization increase. So occur such acts as the swarming of the palolo worm. Another way of boosting odds for success is by dense clustering of organisms in one place—as with barnacles on rocks. The marine worm *Bonellia* takes a different tack. Its larvae scatter; those that settle by themselves become females, those that sink to rest on or in females become minute males and live attached to their spouses. The male of the deep-sea anglerfish goes a bit further; he attaches himself to the female's genital region or elsewhere on her body; there he develops as a parasite with no more structure than that essential for sperm production.

For animals on land, a broadcast release of sex cells would mean their quick drying and death. Internal fertilization is vital, yet more difficulty is involved in thus bringing together the egg and sperm needed for reproductive success. Both body structures and courtship mechanisms have evolved for the purpose.

Great energy goes into courtship. For example, male manakins—small fruit-eating birds of the tropical forests—gather at display areas called leks. Females go there for

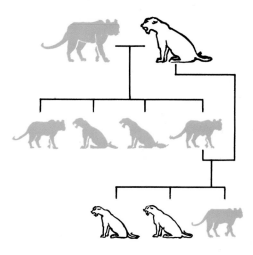

Genes at work in a feline family tree augment the breed of a rare jungle ghost, the white tiger. Accounts of such cats sprinkle hunters' diaries. In 1951 a male was trapped in the forests of India's Rewa district. Rewa's maharaja named him Mohan and mated him with a normal tigress; all their cubs were orange and black. But each cub carried latent—recessive—genes that could produce a white pelt. Mohan's mating with daughter Radha brought them out. Result: such litters as one having two snowy cubs and a third with normal coat (right). From a litter of four partial albinos came a female named Mohini that began the line of white tigers at the National Zoo in Washington, D. C.

302

THOMAS J. ABERCROMBIE, NATIONAL GEOGRAPHIC STAFF.
CHART BY LISA BIGANZOLI, GEOGRAPHIC ART DIVISION

mating. The cock's courtship involves spectacular jumps and dramatic rotations in midair, accompanied by a great variety of mechanical sounds—snaps, cracks, whirrings, buzzes, and clicks. A lek may have from 6 to 60 males. Each stakes out his own perch, stripping the leaves around it to form a sort of arboreal court. Bowerbirds of Australia, New Guinea, and nearby islands also make elaborate display sites.

But complex courting systems are by no means the prerogative of birds alone. An African antelope, the Uganda kob, maintains elaborate leks. Males form arenas in which they prance with head uplifted and forelegs flashing black to entice meandering does. Males which have wrested territories at the lek's center tend to be most successful; they may mate with several females while those at the edge may unite with none. Leaving a site can mean having to win a place all over again.

Natural elimination of ineffective individuals ensures that the stronger, wilier males—and the females they mate with—will pass their gene complements to the next generation. In most mammals the female carries the major burden of rearing young. The male then is free to boost his chances of reproduction by mating with many females. Promiscuity or polygamy therefore occurs more commonly among mammals than birds. For the

latter, the need for cooperation in rearing tends to dampen male enthusiasm for continuous sexual adventure. Thus polygamous species tend to be those in which the females are most able to raise the young alone.

Consider one of my favorite bird families, the weavers. Some species are polygamous, some monogamous. A male village weaver of Africa's grasslands mates with the several females he successfully establishes at his tenement nests. He helps each with her brood to a degree. But in the harsher sub-Saharan environment, where the breeding season is shorter and the food resources less, the related quelea weaver takes only one mate. Both the village weaver and the quelea are seed-eaters. Other species of weavers live on insects in the tropical forests. Here food is plentiful, yet the birds are monogamous. The explanation? Populations are large, breeding seasons long, and competition for food intense. This appears to account both for the help the male must give the female and for the two eggs in their average clutch, compared to the three of their seed-eating savanna relatives.

All birds and mammals, and certain fish, spend much time and energy protecting offspring and rearing them in special ways. Some fish carry their broods about within their mouths; others build nests. Birds' nests come in all sizes, from the tiny cups of hummingbirds to the vast mounds of swans. Forms vary from the simple crossed sticks of pigeon platforms to the elaborately interlaced structures of weavers.

In India, with the start of the monsoon rains, I have watched flocks of weavers descend into their palm tree colonies, change their drab dry-season dress to a plumage of gaudy

With stiff-legged gait and head high to flaunt his snowy throat, an aroused Uganda kob woos a roaming female. The antelope stages his intensely territorial mating behavior in arenas where males stake out individual breeding sites; long grass marks off trampled grounds they rule.

Males of some animal species fight not for ground but right to females. To a kob, territory is paramount, since none mates without it—and those near arena centers mate the most. In ritual battles a male defends his site, locking horns with a rival. A wandering female chooses not so much a mate as his land, and grazes there in a pre-coital rite. Breeding among kobs goes on all year. But a male may fight off others and serve estrous females a few weeks, then run with bachelor herds.

yellows and blacks, and then flutter like glowing blossoms at the lips of woven baskets—attracting willing females from afar. Some species build globular nests supported by bushy shrubs; others shape pendulous nests that hang from branch tips on long tubes. Once I was observing a colony of Baya weavers above a well in an Indian village when a snake came slithering along a limb. It tried to crawl down to the nest opening by spiraling around the nest tube. But the tube's fine mesh contracted, causing the reptile to slip and plunge into the water below. Nature has a purpose to its shaping ways.

With the prime role in parental care among mammals falling usually on females, the mother-litter social unit becomes perhaps the commonest type of group in existence. In relatively few species is the male involved in joint social life with females. African wild dogs may regurgitate part of a kill to feed cubs. Beaver pairs may collaborate in building a dam and den for the security and sustenance they bring to the young. But rarely, except in primates, does this involvement between sexes go many stages further.

I shall never forget the day when, traveling by mule caravan on the desolate mountain pastures of the spectacular high Simen in northern Ethiopia, I first came across the object of one of my studies—the gelada baboon. In the glow of evening sunlight, with wind rippling their long fur, several dozen animals sat grooming or lazily foraging on the grassy edge of an enormous abyss. In the distance, framed by measureless space, the empty landscape rolled away toward the Sudan. The animals were in small groups, each led by a single large richly maned male.

My geladas exemplified a harem-type society in which relatively long-term bonds are established between a male and several females. The herbivorous geladas may form bands 400 strong over rich grassy areas as the rains come; in the long arid season when food grows scarce, harems separate and are spaced farther apart. In other areas where more predators threaten but where food is more plentiful, the troops of other monkey species often contain several males but no harems. The additional males lend protection, and the food supply permits exploitation by both sexes. But when many males are present their conflict results in formation of complex dominance hierarchies. Often coteries of friends support one another; the top, or "alpha," male holds his position by having henchmen that gang up to oust a challenger.

Animal societies have developed from simple congregations that join for mating into complex social systems in which adults exhibit cooperative and even altruistic behavior so that their young can grow in safety and acquire the skills of adult life. In such evolution there are numerous prefigurements of the even more complex pattern of human life. For man, perhaps the most fundamental problem of our time is an adjustment between competition and cooperation at every level of human function—from personal through familial to international. An understanding of these processes in higher mammals may provide us with an account of the sources from which human cooperative behavior springs—and of the forces that prevent its expression.

A bushbaby mother and her young cling to the branches of a shadowy world. Primitive primates, these bushbabies are superbly adapted for a nocturnal life. Big eyes scan the night. Long hind legs propel adults on broad jumps of 16 feet. Pads on toes aid in grasping. And, tuned to the rhythm of life, the mother commonly gives birth to twins twice a year—when food abounds for her babies of the bush.

THEO BEKKER

Wolfgang Wickler

COURTSHIP IN A WATERY REALM

G radually the little male stickleback in the aquarium changed color while I watched. The barred back that helped camouflage him among the weedy growths became a flashy blue. Pale undersides blushed, then intensified to a vivid red. And once his turnabout tints reached their brightest, he began to patrol his territory with darting moves. This showiness added to the conspicuousness of his dress. Fortunately for him, I thought, no predator lurked to be drawn by the jaunty display.

Why should a stickleback invite danger by calling attention to himself? The answer: courtship—that complicated system of signal and response sparked by the urge to perpetuate the species.

Among mammals, sex rites often are circumspect, even perfunctory. But in the realms of birds and underwater creatures, courtship follows elaborate patterns. It serves three vital functions: attracting a mate, minimizing hybridization through a system that beckons a mate of exactly the right species, and prodding both partners to simultaneous sexual arousal.

Rites of courtship and breeding behavior in the watery world are as varied as any on land. The soft-bodied squid flashes color and waves lithe arms before transferring packets of sperm to the female; she plants her pods of jelly-encased eggs and jets away, leaving the young to face the world alone. The less primitive stickleback builds a nest, mates with several females, then assumes the chore of caring for eggs and brood. The even more complex discus fish shares duties in raising fry, nursing with a skin secretion the hatchlings that cling to his and his mate's sides.

There are more fishes on earth—in number of species as well as individuals—than all the mammals, birds, reptiles, and amphibians

Sinuous arms of a spawning squid add an egg pod to strands in a communal nursery; many females may join to make mop-like clumps several feet across— one of the wondrous rites of reproduction among underwater creatures.

ROBERT F. SISSON, NATIONAL GEOGRAPHIC PHOTOGRAPHER

Spots before his eyes on fishes in a laboratory tank foretell for author Wickler a change to reproductive behavior. Normal marks are the vertical stripes on the fish nearest the bottom of the tank; all are Tilapia, *fishes of the cichlid family.*

Fishes shift color patterns by spreading or concentrating pigments that lie in skin cells. Some changes can be effected in seconds, others take days.

Scientists theorize that color change evolved for protection, then found other behavioral uses. Studies with models gave clues to color's import. Red undersides on a roughly shaped bit of wood, for example, stirred males of some species to attack; they ignored a finely carved, but plain-bellied, dummy.

Orange dots on the anal fin of a Haplochromis *(model above) simulate spawned eggs. A male fish uses such spots to tempt a female into trying to scoop them up along with her eggs; instead she mouths his milt. Eggs and hatched brood stay in her mouth until young can swim.*

Mouthbrooders often guard fry from danger by letting them dart back into parental jaws, as do these Tilapia *(opposite, lower). In some fish species males do the rearing. About 200 kinds of fishes are mouthbrooders—all in waters thick with predators but thin with havens for eggs or fry.*

OPPOSITE, UPPER: GEORGE F. MOBLEY, NATIONAL GEOGRAPHIC PHOTOGRAPHER. LOWER: RAIMUND APFELBACH

taken together. Not surprisingly, among fishes we find a wide range of social and breeding behavior. One family in particular, the cichlids, offers a fascinating array.

Cichlids are perchlike tropical freshwater fishes. Several hundred known kinds inhabit the waters of Africa and Central and South America. Most are small—seldom more than a foot long. Some are food fishes; *Tilapia* species, prolific and fast-growing—50 can become 3,500 one-pounders in little more than a year—have been widely introduced as a protein source, especially in underdeveloped nations. Scores of other species of the hardy and pugnacious cichlids adorn hobbyists' aquariums. Such names as African jewelfish, veiled angel, firemouth, orange chromide, and Egyptian mouthbrooder attest to their beauty and interesting ways.

To the nutritionist and hobbyist, cichlids provide a mine of reward. For the biologist, the lode is even richer. Encounters with the unexpected in their structure and behavior are ever intriguing. For me, cichlids have proved an absorbing 14-year study.

About five million years ago earth's crust was torn asunder, creating Africa's Great Rift Valley, one of the globe's most awesome fractures. Rivers were stopped, or their flow reversed. Enormous basins opened and filled with water, forming the Red Sea and some of the world's largest lakes. Today's population of fishes in those lakes derives from the stock that once inhabited the rivers. In the same way that the varied Darwin finches evolved on the isolated Galapagos Islands (page 185), many surprising species of cichlids evolved in these isolated African waters.

Especially in Lakes Tanganyika and Nyasa cichlids underwent explosive speciation—rapid evolution of many closely related species in a circumscribed area. Each lake has species found nowhere else, exploiting virtually every habitat and way of life.

There are cichlids which school just off shore and feed on plankton. Some crack shells of mollusks to eat the fleshy contents. Some crop vegetation like aquatic cows. Others rasp minute plants from rock surfaces or pick tiny insects and crustacea from crevices. Everywhere predatory species prey on smaller kin. There are kinds which specialize in devouring young cichlids, or which feed on scales they scrape off their neighbors. One species dines on small organisms but prefers biting off the eyes of carp-like fish.

Cichlids likewise have developed diverse ways of caring for offspring. They form troops, harems, and every possible kind of family —ones in which the mother alone rears the young, ones with the father performing that chore, ones with both adults sharing. And they show a remarkable variety of reproductive habits.

Clues from fossils suggest that cichlids originally reproduced in monogamous families. Such species as the African jewelfish retain that trait today. Male and female team to scrub algae from a flat rock, where eggs are then spawned and fertilized. They take turns fanning with their fins to aerate the eggs. When the young hatch, the parents scoop them into their mouths, taking them each day to different protective pits previously dug in the sand.

Monogamous male and female cichlids look alike and perform family chores interchangeably. But the slightly smaller female tends to stay closer and longer with eggs and young; the male tends to defend territorial boundaries against intruders. Some such difference in the distant past gave rise to the marked division of parental labor now exemplified by cave-brooding cichlids, and also to the evolu-

tionary lines that led away from monogamy and resulted in today's many species of mouthbrooders and harem formers.

Hiding eggs in caverns or holes becomes an effective means of protection; smaller nest holes with narrow entrances are easier to defend, but have room for only one parent. Thus cave-brooders specialized, the male for guarding, the female for tending offspring. Freed from brood chores, the male could hold an area for several females and use his full reproductive capacity in a harem system.

Lamprologus congolensis is a small bottom-living cichlid from the Congo region. The male sets up a territory in a place where he can dig nest holes under stones and other objects. He courts until a female follows him to residence in one of his nests. Then he goes wooing again. Should he return with a second female, the first will

Watery "push of war" pits male cichlids in territorial battle. Disputes between neighbors over boundaries usually call forth bluffing displays. But when one fish enters an area claimed by another, fights follow a sterner ritual. Males circle each other, trying to butt the opponent's flank. Tails thrash to send surges of water against a rival— their pressure a measure of the maker's strength. Such tactics may settle an encounter, one fish yielding and swimming away.

Some fights, however, climax in mouth-to-mouth dueling. Among cichlids that lay eggs on smooth surfaces or underwater vegetation, biting and tugging is the rule; the fish getting a hold on a foe's upper jaw normally wins. But mouthbrooding cichlids such as these battlers in an aquarium usually just shove with their jaws agape.

RAIMUND APFELBACH

The stickleback: zigzag path to bachelor fatherhood

Pugnacious denizens of Northern Hemisphere coasts, ponds, and streams, sticklebacks taught ethologists much about how signals and responses regulate animal behavior. One species, the three-spined, schools in drab stripes. But in spring males swim alone, begin a change to breeding dress, and vie for territories. They flaunt red bellies, dig snouts into the sand in threat displays.

When one bluffs off his rivals, he builds a nest of algae—glued with a kidney secretion—then pushes a tunnel through. Now in his gaudiest hues, he attracts an egg-swollen female. He courts her with a zigzag dance; if she accepts, she signals by a head-up posture. Immediately he leads her to the nest, where he lies on his side to point out the entrance. She wriggles in, and his quivering touch prompts her to spawn. In turn he enters and fertilizes the eggs—then chases her away; she plays no role in rearing young.

The male may entice several mates to the nest. Afterward he hovers at the entrance, fanning his fins so that a current keeps oxygen supplied to the eggs. They hatch, and he guards the brood, seizing errant fry in his mouth and spitting them back into the swarm. Dull dress returns when drab color cells expand to mask the bright ones beneath.

PAINTINGS BY RICHARD SCHLECHT

attempt to chase her away. Now the male must drive his number-one mate into her hole and keep her there until number two is established nearby. Bringing home a third mate busies him with a contentious two—plus fighting off rival males that may covet one of his harem. Seldom does a *Lamprologus* male have the aggressive capacity to handle more than three females.

Monogamous cichlids form pair bonds and recognize each other individually. But the polygamous *Lamprologus* apparently has no such ken. Male and female seem bonded not toward each other but independently to the same breeding area.

Conditions that favor polygamy have other effects. As long as rearing of the young takes place in the male's territory, his reproductive success will depend directly on how many females the territory can accommodate at one time. The number of females is influenced by territory size, which is related to the size of its owner and how well he defends it. The genetic result of such a system is the development of very large males—which in some cichlid species are twice as large as the females.

When young are not reared in the male's territory, polygamous conditions may lead to mouthbrooding. Now the male's ground is merely a spawning site. He waits for ready-to-spawn females to come by, courting one at a time and leading each to a spawning pit where she lays her eggs. When these are fertilized, she takes them into her mouth and swims away to rear the brood alone.

Among primitive mouthbrooders, the female, prior to spawning, stores up fat on which she lives while her mouth is blocked with eggs. But more specialized mouthbrooders have developed the ability to feed without swallowing the eggs; they can brood larger eggs for a longer time—43 days for one species.

Eggs of mouthbrooders may be as large as peas, and from 10 to 20 in number. Monogamous cichlids, which glue their eggs to plants or smooth surfaces, spawn up to 1,000 eggs pinhead-size or smaller. By contrast, herring, carp, salmon, and other fishes produce tens of thousands of eggs, deposited among aquatic vegetation or buried in the sandy bottom. But in the watery world of no quarter, most of the eggs and newly hatched larvae die or are eaten. Even among species that produce great numbers of eggs, on the average only one of the young survives to replace each parent fish.

The fewer eggs, the more valuable each is to the species. As mouthbrooders become more specialized, the interval between spawning and take-up becomes shorter, thus lessening the time that the eggs are exposed to predators. In one genus, *Haplochromis,* the female snaps them up immediately. This leaves no time for

fertilization—a catastrophic consequence except for a bit of deception by the male, a practice which I discovered a few years ago.

Males of this genus have conspicuous orange or reddish spots on their anal fins. At mating time the color intensifies and the spots mimic the female's eggs both in size and hue. After the female spawns, the male drags his spread anal fin over the sand, simultaneously emitting milt. The female, alert for any eggs she may have overlooked, tries to snap up the male's egg spots—and in doing so gets a mouthful of fertilizing milt.

Haplochromis males live in colonies and compete for every female that comes by. The most attractive males have the best reproductive success. A courting male poses in front of the female, quivers with spread fins, and shows his nuptial colors.

In cichlids, as in other fish species, color depends on pigment particles in special cells—chromatophores—distributed over the body surface. These flat cells reach out in rootlike branchings. Nerve impulses may cause the color particles to spread through the cell, suffusing it with red, or yellow, or black, for example. Or the pigment may concentrate in the center, leaving most of the cell pale. Changes in actions and mood are accompanied by changes in color patterns. Thus an experienced observer can predict what a fish is going to do by looking at its coloration.

To become experienced means logging countless hours before an aquarium. And solving tantalizing puzzles, I have found.

In the waters of Lake Tanganyika, and only there, swims a cichlid called *Tropheus moorei*. Normally the fish is nearly all black, but it

Trailing draperied fins and spreading crimson-tipped gill covers, a Siamese fighting fish woos a female in a chase punctuated by courting poses. This easy-to-raise aquarium favorite—varieties come in a rainbow of colors—leads his mate to a bubble nest built by trapping gulps of air in sticky mucus. Then the male snaps up the sinking eggs and spits them into the nest; there he will tend them while embryos develop.
ARLAN R. WIKER

can display a pale gray when frightened, a few vertical yellow stripes when angry, and a broad golden band down the flanks in certain other situations. Therein arose my puzzle.

Since the male *Tropheus* does not cooperate in brood care, one would expect marked sexual dimorphism in this highly specialized mouthbrooding species. Yet the sexes are so remarkably alike that even experienced cichlidologists have difficulty telling them apart. During courtship the male displays bright gold from dorsal fin to belly, making him look much like a glowing coal. But as I watched *Tropheus* in a tank, two strange facts assailed me. First, every fish from time to time displayed the gold girdle and did a quiver dance without any other sexual behavior. Second, the fish formed neither monogamous pairs nor harems, but swam in groups with no indication of territoriality.

In time the explanation came. *Tropheus moorei* lives in "closed groups"—tight coteries in which members of each group know one another individually. They tolerate the others of their clique but viciously attack any outsider. A fish separated from his own group by a glass partition will desperately try to rejoin it, and will rarely seek alliance with—or win acceptance by—another group.

Truculent *Tropheus* often behaves combatively even toward members of its own coterie. But in a confrontation the attacked fish stops the aggressor by starting a quiver and girdle display. It happens regardless of the sex of the two fishes. Thus a sexual signal serves also as an appeasement gesture; it elicits sexual responses in the attacker—usually not strong enough to trigger sex behavior but enough to counteract aggression. What was at first a courtship device becomes a signal of broader social meaning.

Other animals reflect such processes. Baboons and other Old World monkeys, for example, use pseudo-sexual behavior in much the same way to pacify attacking group members.

Cichlids, then, have evolved a variety of social arrangements, from the monogamous family to the closed group comparable in structure to baboon troops (page 392). These remarkable fishes, as a result, provide opportunities for studying the derivation of those animal societies. Equally important, they offer experimental tools for better understanding the evolution, function, and ecological implications of animal social behavior in general.

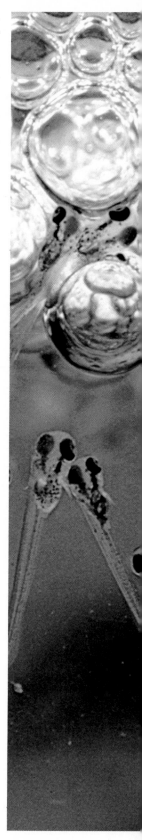

Popeyed slivers, newly hatched Siamese fighting fish hover beneath their frothy nest. If they drift toward hazardous deeper waters, the attendant father will nudge them back. But after a few days, when they are able to swim freely, they must hide in weedy shallows or he may eat them. Such behavior in a world rarely seen by man rivals the familiar dramas of life on land.

ARLAN R. WIKER

Dale F. Lott

THE WAY OF THE BISON:
FIGHTING TO DOMINATE

A bull rolls hard in the loose dust of the wallow, slamming his body sideways and flailing his legs in the air. Dust explodes from beneath him and envelops him. Now he gets up and heads toward another bull a hundred feet away. He lifts his muzzle, opens his mouth, and lets out a deafening bellow. My teen-age son shifts beside me in the seat of the jeep, here on the National Bison Range in Montana, and the awe that I am feeling I hear in Terry's voice: "He's *really* enraged."

The bull stalks toward his opponent, moving in stiff little steps, banging his front feet down so hard that the shaggy "pantaloons" on his front legs dance and his hoofs ring when he crosses hard ground. Now he stops bellowing and exhales explosively through his nostrils in rhythm to the stamping of his front feet. His tail stands up like a living question mark.

Bison bulls are incredibly strong. In a charge they can shatter planks two inches thick and a foot wide. Once I saw a bull put his 2,000 pounds and the strength of his neck behind his horns and in seconds rip through the 3/4-inch plywood facing of a corral gate.

The bull now being challenged has been on the receiving end of that power. An 18-inch scar runs up his ribs; his horn tips, shattered in other battles, are blunted and worn. I fumble in my pocket for my little tape recorder to describe the scene, and fleetingly I wonder what I would do in his place.

He doesn't yield a step. He scrapes the soil with one foot, flinging dust up onto his belly. His back arches as he answers the oncoming bull bellow for bellow. They close, walking stiffly, heads held low and slightly to one side until they almost meet. Then their heads swing sharply into line with their bodies, their muzzles tuck down and back, and they slam together.

The impact of their foreheads makes a surprisingly soft sound, cushioned as they are by several inches of curly hair between their

Bellowing ton of unpredictable fury, a bison bull looms out of a dust cloud on the National Bison Range in Montana. During the summer breeding season he will join dozens of other bulls as they square off in spectacular "fighting storms" that brew new dominance relationships. Despite his weight and size—9 to 12 feet long and 5 or 6 feet from hoof to hump—a bull can run at an estimated 30 miles an hour and wheel with amazing agility. Bison bones have been found on mountain heights where horses could not climb.

Commonly called buffalo, bison are more closely related to the wisent of Europe than to buffaloes of Asia and Africa.
LOWELL GEORGIA

Closing in on a cow, an older bull (center) has little to fear from the eager 4-year-old at right. Males

LOWELL GEORGIA

...usually don't breed until their fifth year, when they are strong enough to dominate other mature bulls.

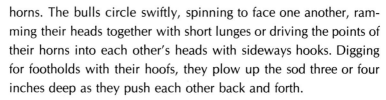

horns. The bulls circle swiftly, spinning to face one another, ramming their heads together with short lunges or driving the points of their horns into each other's heads with sideways hooks. Digging for footholds with their hoofs, they plow up the sod three or four inches deep as they push each other back and forth.

Suddenly, with 40 seconds on the stopwatch—a long fight—it's over. One of the bulls has turned his head and neck away from the other—part of the submission movement of turning and fleeing—and both abruptly stop. They stand quietly for a few seconds and then walk peacefully away from each other.

"I'll take the one on the left," I say quickly. As the dust settles, we scramble to identify the bulls. With spotting scopes, binoculars, and the naked eye, we search for distinguishing features. "Where's the buffalo book?" Terry asks. Under something. We dig through the disorganization of our little mobile office. Finally we find our precious notebook of photographs, sketches, and scribbles that record the hump shape, horns, beard, scars, brands, and other marks that identify each of the 35 mature bulls in this area of the range. When we check off these two as Rib Rip and Black Beret, I dictate the field notes and we relax.

We have taped the outcome of an interaction. It is only one piece of data among hundreds that we will record this summer about the 400 or so bison that roam the range in the Flathead Valley. But that piece helps us learn something more about the dominance relationship of these bulls. And it is a satisfying accomplishment. Now the August midmorning sun doesn't seem nearly as hot as it did five minutes before, the churned-up dust that hangs over the breeding herd isn't as choking, and even the dry brown grass that covers the gentle hills seems more alive and inviting.

I have known these hills all my life; for much of my childhood my parents lived on or near the range. When I finished my formal training in comparative pyschology, bison seemed a natural subject for field work. They were prodigiously successful animals once, roaming North America in the millions, and I wanted to learn how their social behavior contributed to that success.

The breeding season, or rut—late July to the middle of August—is the major social event of the year. At all other times the mature bulls (those five years and older) live either alone or in small groups.

Wallowing in the dirt, a bull raises a powdery veil behind the cow he is tending. His dusty performance may be a threat display aimed at challenging bulls. "Testing" a cow, a bull reacts by lifting his head and extending his muzzle in flehmen (left), a posture triggered by smell. Bulls, accustomed to wheeled intruders, sound off close enough to the author's jeep for a drive-in taping.
LOWELL GEORGIA

Occasionally they visit a cow herd, which usually consists of 10 to 50 cows, calves, and young bulls. But during the rut they stay in the cow herd most of the time. And while they are in the herd nothing matters so much to them then as their relationships to each other. This so preoccupies them that they seldom eat. Sometimes they even abandon a receptive cow to seek out a distant challenger and try to dominate him.

The primary value of being dominant, presumably, is to be able to stay with cows in estrus and breed them. In the natural selection process, the breeding itself spells success. But that is not the major motivation of the bulls. They want to dominate other males.

Fighting is one way that dominance relationships can be established and maintained. Because fighting is dangerous and demands so much time and energy, substitutes have developed. In animals that establish—and defend—territories (page 230), fighting is often avoided because individuals are separated by distance. But species whose social life is organized by dominance depend heavily upon the ability to predict each other's behavior from such signals as postures and vocalizations.

Many of these signals are obvious in bison. The snorting, bellowing, stamping, head-on approach is easy to read. As I continue to watch the animals, I learn to discern less obvious signals. At least once a season I see one more reason to shake my head in wonder at the subtlety of the big brutes.

Take the broadside threat, for example. Like the male of many related species, the bison bull sometimes threatens by turning exactly broadside to his opponent and drawing his body stiffly up to maximum size. In full bellowing display, this pattern is easy to spot. So is the full response of the submitting bull: He swings his head and neck widely to one side and sometimes grazes in a ceremonial sign of peace. But often both signals are sent in the most fleeting form. One bull walking broadside to another may stiffen for half a stride in a momentary threat, while the other points his muzzle a bare three inches to one side in submission. The signals are sent and received in a fraction of a second.

The most rigid form of dominance organization is the linear hierarchy, where the order forms a straight line. The top-ranking individual dominates all the others and the bottom-ranking submits to all those above. Chickens have such a social order (page 34).

To learn about dominance in bison, we recorded the outcome of several hundred encounters between 35 mature bulls during one breeding season. Though they did exhibit some linear dominance, they also showed triangular relationships—A dominates B,

Status-seeking bulls collide head on in a duel for dominance. Males also battle for possession of cows in estrus, the winner tending his prize by standing closely beside her (opposite). To keep her from wandering— and becoming fair game for rivals like those circling nearby —he will step in front of her every time she makes a move.

A bison bull engages in little or no erotic play preliminary to mating, though he may pant and swing his chin up and toward the cow's back. He sometimes uses this "intention" movement as a response display toward a bull that is submitting.

LOWELL GEORGIA

Springtime in the Rockies finds cows, nine months after the rut, in calving herds in the blossom-strewn

Flathead Valley that serves as Dr. Lott's outdoor laboratory. Cows begin breeding when 2 years old.

B dominates *C*, but *C* dominates *A*—as shown in this typical episode drawn from my notes (elapsed time about 15 minutes):

Groin Scar displaces Widower beside Miss Universe. Widower turns broadside, walks away, and grazes. Groin Scar displaced by Flat Top without protest, walking out about 20 feet. Widower comes back and displaces Flat Top. Flat Top walks right past Groin Scar, who immediately returns and displaces Widower beside Miss Universe. Flat Top displaces Groin Scar and the cycle is repeated. No resistance by any of the bulls.

Linear or triangular, dominance relationships are not necessarily permanent. Subordinates may turn the tables on former superiors. In my study I found that 12 percent of the victories were reversals. A dominance system is advantageous because it reduces fighting and affords social stability. Ironically, the stability is advantageous to all only if it is temporary. For if there were never any changes in status, subordinate bulls would never have opportunities to breed.

"Fighting storms" seem to meet this need. Every few days during the rut, fighting breaks out for no observable reason. No bull will yield to another; every challenge is met with a charge. Within two hours 50 or 60 fights may erupt among the bulls. Relationships are re-arranged, and some subordinates become dominants. Now they too will get the chance to breed and pass on their genes.

Subtle though he may be with another male, a bull is quite direct when he approaches a cow. There is nothing that deserves the name of courtship. He "tests" her, sniffing and licking, then extends his muzzle and lifts his head in a posture called *flehmen*, practiced by other ungulates (page 242). In this way he probably learns something about the sexual receptivity of the cow, but the accuracy of his chemical analysis does not rate very high.

The bull may move on. If he stays, he tends the cow, standing head-and-head beside her, patiently waiting, threatening or fighting bulls that challenge him. He tries repeatedly to mount her, but she steps aside each time. Finally she stands, union is completed in seconds, and bison life is perpetuated.

She will bear the new life through the winter. Then, when May comes to the range, she will be one of many cows, heavy with young, that cluster in the lush green grass. One day she heaves to her feet and walks away from the others. She seems nervous. She stops and grazes fitfully, lying down . . . getting up . . . lying down. After an hour or two she gives birth to her calf and licks it clean. Within half an hour the calf gets to its feet, finds one of her legs, and follows it up to her belly and a teat for its first meal. Later—maybe

Wide-eyed mother in partially shed winter coat bawls a warning to stay away from her calf. Though not as large as bulls, cows can be dangerous. In defense of their calves, cows will attack young bulls, even charge men on horseback.

A cow usually leaves the herd to give birth; licking the calf, she imprints its sight, smell, and taste. The 25- to 40-pound calf, able within 30 minutes to stand and suckle, trails its mother back to the herd. Born late April to June, a calf will continue to nurse until winter, when its mother weans it by rejecting its nuzzling.

Though somewhat similar to domestic cattle in behavior and anatomy, bison have 14 pairs of ribs instead of 13. In their prime between 6 and 15 years, they sometimes live to 30.
LOWELL GEORGIA

an hour and a half or a day and a half—the cow returns with her calf, ending her longest separation from the herd since the birth of her previous calf the season before.

Bison evolved into a gregarious society for protection against predators like the wolf. Why then did this cow leave the herd just when she was so vulnerable? To learn, apparently, something as important to her as life itself: which calf is hers. Away from the herd she could get to know her calf, sniffing and nuzzling and licking it, imprinting it upon her mind and senses.

This cow-calf relationship can be established in the herd; not all cows isolate themselves at calving time. But those that do not leave the herd risk a mix-up. For perhaps an hour after giving birth, a cow will accept the first calf that comes in contact with her. She will reject all others. Newborn calves will approach any cow—or any large animal, even a horse. So the cow that gives birth in the closeness of the herd may mistakenly adopt another's calf.

As the months pass and another rutting season approaches, the calves' coats, yellowish-red at birth, are turning dark brown and tiny horns are sprouting on their heads. They wander away from their mothers to romp with other calves. They butt heads in play, exhibiting the vigor and quickness of their parents. In five or six years the males will be as large as their fathers. We're sometimes told that the competition in which the fathers won breeding rights selected the winners as the best bulls, in the sense of "all-around best." The fact is, there are some things they just can't do well.

I remember the day I watched three mature bulls and a younger one chase a cow. For the first 200 yards the older bulls ran easily, their long hair flowing in the wind. They even attempted to mount the cow at a full gallop. But five minutes later the cow and the younger bull circled back, still at a full gallop while the mature bulls followed at a wobbly walk, tongues hanging out, sides heaving. They were just too big to run very far.

This deficiency is part of the compromise of natural selection, which gave them the one trait that makes them successful specialists: the power to force other bulls aside at tending time. It is then—when the bull moves to his task, beard swaying and pantaloons jouncing, belly lifted in an intense arch as he bellows out a challenge—that he is perfect. And he is magnificent.

Brave bulls of tomorrow, 4-month-old calves rehearse for serious combat. Playful as youngsters, they will grow progressively wary and dangerous, especially to a man approaching on foot. They hone their fighting skills for those days of the rut when to be a bull is to dominate other males, to claim a receptive cow—and to sire another generation of a rugged breed.

LOWELL GEORGIA

Richard L. Penney

POLAR RENDEZVOUS
IN A PENGUIN
COLONY

t is minus 20° Fahrenheit in the austral springtime and for yet another day the wind is rattling the hut's stovepipe. Soon the penguins will be returning to these Antarctic shores. After breakfast we don boots and parkas and trudge down through the barren, guano-laden areas of Cape Crozier on Ross Island, a necessary prelude to the excitement of seeing the first arrivals of the breeding season waddle across the ice from the open sea. Or—even more important —of welcoming back some of "our" penguins, banded birds of known history, age, and sex.

For many years we have banded penguins to identify them individually. By closely observing the birds and the nests that we have marked and mapped, we can study basic behavior and population changes. Of the 18 penguin species, the Adelie is the southernmost,

Belly-flopping off an ice cliff, breeding Adelie penguins forage in frigid polar seas for themselves and their young. Lively and gregarious, the 14-inch Adelies winter on the floating belt of pack ice surrounding Antarctica, homing in spring to vast rookeries on its rocky coasts and islands. Here, enduring hunger, harsh winds, and cold, they build nests, mate, and raise their chicks.

Flightless like all penguins, Adelies possess such stamina and navigational ability that birds experimentally displaced by the author have walked, tobogganed on their bellies, and swum as far as 4,000 miles to reach their natal ground.
DAVID H. THOMPSON

335

Summer population crowds a beach at Cape Crozier. Amid shattering calls, audible a mile away, 300,000 Adelie penguins claim rookery space, uniting with previous mates or seeking new ones. Unlike the larger emperor penguin, which breeds in winter and has no nest, the Adelie is a highly territorial species. Nest plots in the compact colony average about eight square feet. Annual banding of chicks (above) helps biologists study population dynamics and behavior by age.

ROBERT W. MADDEN
OPPOSITE: W. J. L. SLADEN

breeding within 800 miles of the Pole. It is also possibly the best known and most studied. As long ago as the late 1800's a French museum curator in effect pronounced man's knowledge of the Adelie penguin complete. He was wrong. Man still is learning.

This is a summary of six seasons of work at several Antarctic locations. You might also consider it a composite of my experiences and those of others over a 14-year period. My own research was sponsored by the National Science Foundation.

Field work on earth's coldest, windiest continent calls for a special know-how and equipment ranging from ski-equipped aircraft to long underwear. At the onset of spring—late September— the penguinologist brings in food and apparatus to remote rookeries either by dog sledge (in my earlier studies) or by helicopter. Adelie penguins usually return in late October, the exact time depending on weather and ice conditions. The only way to be on hand to greet the first arrivals is to walk down to the beach.

This particular morning at Cape Crozier, as I wandered along the edge of the ice foot (ice firmly frozen to the shoreline), I heard a peculiar sound. It varied in pitch and came from underwater, possibly from beneath the overhanging ice foot. I could see no penguins. Other than the wind, all was silence. In an effort to discover the source of the unfamiliar sound, but without any great logic, I took out my field knife and chipped off a piece of ice, tossing it into the open lead of water adjacent to the ice foot. Near the spot where it splashed a leopard seal surfaced, then dived. We biologists were not the only reception committee awaiting the Adelies!

Long fascinated by this streamlined predator of south polar waters, I yelled up to a friend mapping penguin nests in the rookery above, "Bob, I've found a way to attract leopard seals."

In went another piece of ice. Out came the seal. He surfaced, cruised toward us, and dived close to the ice foot. I leaned over the edge to watch through the clear blue water. Up from the depths he streaked and, mouth wide open, made a six-foot lunge out of the water directly at me. Automatically I jumped back, extending my knife to protect myself. The seal bit down and slipped back into the water, leaving some flesh on the sub-zero knifeblade.

This was one of several instances when I have been chased by leopard seals early in the season. Evidently, before food becomes plentiful, these hunters do not differentiate between two-legged penguins, their normal prey, and two-legged penguinologists.

Within a few days our springtime vigil is rewarded. Black dots appear on the horizon—the first penguins. When the snow cover on the sea ice is soft and deep, the Adelies advance single file,

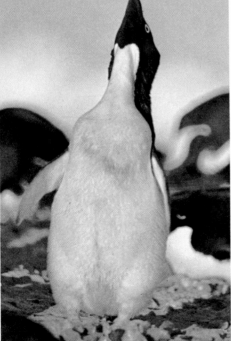

formed in delightful trains with various individuals taking turns as engine, breaking trail. While all are fun to watch, the field biologist looks particularly for the bird with a band on one of its flippers. Through numbered bands we trace previous behavior of the individual, its sex, age, faithfulness to mate and territory, navigational ability, or other data that may be on its record. Marking large numbers of birds permits systematic study of the behavior, physiology, and ecology of the species. At my small rookery at Wilkes Station (2,700 adult penguins) spotting banded individuals was relatively easy. But what a formidable task to find a particular bird among Cape Crozier's 300,000 — and all look alike to the last feather!

As the influx reaches its peak, about nine days later, the Adelie rookery becomes a hullabaloo. Despite the noise and confusion, the behavior of breeding adults follows a highly organized and synchronized pattern. Interactions of the pairs are quite predictable. Males generally return first. Those that have bred before claim the nest sites they held the previous year. Giving a loud *gaauaauaaah,* they raise their beaks skyward, roll their eyes down, and wave their flippers in an ecstatic display attractive to females.

When the rookery is crowded, a male can easily make a mistake about which piece of rocky ground he had last year. Claim-jumping, or even simple trespass, can bring on serious fights between males. Late arrivals whose sites lie near the center of the rookery must run the gantlet of males already established on the outer sites. Pandemonium prevails for a while. But when the territorial issues are resolved, the penguins settle down to building their nests of stones. The female of a pair assists in the task when she arrives.

Females, like males, return to the previous breeding spot. It is an intense occasion when previously mated birds meet after seven months of separation — something like a penguinologist's coming home from 27 months' isolated field work in Antarctica. A behavioral pattern called the loud mutual display, which is used in a variety of social situations, marks the birds' reunion. They face each other, bills pointing upward, eyes cast downward. The feathers on neck and crest are erect. Both birds press their flippers to their sides and wave their heads from side to side out of phase, uttering a sequence of loud, harsh calls.

At a reunion of mates on their former nesting territory, their distinctive sounds, or calls, help them identify each other. Copulation may occur within minutes. Nest-building continues. Can you tell a male penguin from a female by who presents stones to whom? A charming notion — but inaccurate. Both do it.

Previously mated birds exhibit a marked degree of faithfulness

Sociable at sea, belligerent on the breeding ground, adult birds jealously defend their special nesting site, which they reclaim each spring. A third bird barging in on a wooing pair may be pecked and pummeled as well as scolded.

White eye rings distinguish Adelies from other penguins. Male and female are identical in plumage and body form. But scientists can usually tell which is which by watching social interactions of pairs. The neck-stretching ecstatic display of a male advertises for a mate; female's appeasing bow signals acceptance. They take turns collecting small stones; one stands guard to keep other penguins from stealing them. When the pile is big enough, the birds lie prone and, using their feet, scoop out a hollow where two eggs will be laid.

DAVID H. THOMPSON
LOWER LEFT: ROBERT W. MADDEN

Noisy nuptials have ended. A bright windless day in early summer finds males minding the eggs. Thoug

ind and 24-hour polar sunlight have cleared the colony slope, Cape Hallett remains locked in winter ice.

Down to the sea, grouped for safety, go penguins seeking food after a fast enforced by sitting on the nest. Like most Antarctic animals, they feed only in the oceans, teeming with life in summer and closer now that shore ice has broken away. Stroking with their flippers—modified wings—they cover long distances at ten miles an hour by "porpoising" (opposite), surfacing periodically for air. Underwater, penguins' brief spurts of speed can exceed 30 miles an hour.

to each other. At Wilkes, 84 percent of the adults that had bred one year and returned the next kept the same mates. Data indicate that birds remated with previous partners make more efficient use of the short season and are more successful breeders than newly formed pairs. When I am asked the intimate question of "divorce" rate, I reply 16 percent for Adelies at Wilkes—about half the rate for human beings in the United States.

Naturally, deaths during the winter split some mated pairs. The survivor seeks a new union. A male's ecstatic display attracts any female lacking a mate. She responds by giving an appeasing bow with her head turned to one side, and the male bows in return. These movements are extremely slow and guarded—you might call them Victorian. Penguin courtship loses some of its courtliness,

however, when more than one female answers the male's display with the appropriate bow. Since he will accept either, the females battle it out. We learn a lesson here about animal behavior and survival. A male assigns his highest priority to defending his territory; a female to holding her mate. Rival Adelie females fight each other as furiously as do the males.

Courtship is serious business. After many days of association, loud and quiet mutual displays, repetitions of the ecstatic display, nest-building, and copulation, a pair bond becomes firmly established. In the encircling nest of loose stones the female lays a clutch of two bright greenish-white eggs—and promptly goes to sea to eat and recuperate. The male stays behind to incubate the eggs for the first two weeks of a necessary five.

Life is not easy in Antarctica for man or beast. Penguins depend wholly on the sea for food. Their stubby wings, covered evenly with waterproof feathers as dense as fur, reflect an adaptation through the ages to their marine environment. Speed in the sea, where they "fly" with their flippers, has proven more useful to their survival than flight in the air.

The Adelie's primary food is krill, small shrimplike organisms which the great baleen whales also feed upon. Other crustaceans, known as amphipods and copepods, and small fish (about 30 percent) make up the rest of the Adelie diet. For other species of penguin the food might be anchovies, squid, fish, or various crustaceans. No one has yet made a comparative food study for all species—their range covers nearly two-thirds of the earth's southern seas, stretching from Antarctica nearly to the Equator. One feature all penguin habitats have in common is the presence of cold ocean currents, but not necessarily ice and snow.

The Adelie's food-gathering journey is a hazardous trek indeed. Just how hazardous we have learned, in part, from watching predation by the leopard seal. Along the beaches of Cape Crozier five or six seals kill and eat about 15,000 penguins each season—5 percent of the Adelies' breeding population.

Normally the seals cruise back and forth 10 to 50 yards off the ice foot or beach. Their attention is toward the sea as food-laden adults return to the rookery. As a group of "porpoising" penguins approaches, the seal dives. I don't know what happens underwater; we next see the leopard surface with a penguin in its mouth. The seal tosses its head violently, literally shaking the victim out of its skin and feathers. Biting meticulously, the seal shreds the carcass until all that remains is the head, backbone, and legs.

When female Adelies return to relieve their mates from the nest, an elaborate ceremony ensues. Loud calls between mates reaffirm their identity. As the incubating male gets off the eggs, he often

appears stiff in the joints; he has been sitting tight for two full weeks. When you add to that the time since he arrived at the rookery, you find he has been busy—and without food—for about 35 days. The environment places such severe demands on many colonial pelagic birds.

Before the male departs after his long vigil, he performs one final duty: collecting more stones, which the female incorporates into the nest. A well-made nest kept in constant repair is essential to protect the eggs against wind, meltwater, and accidental displacement. Early in the season drifting snow can bury nest, eggs, and incubating parent. The stones added later help keep the nest above the high-water line in December and January, when thaw streams develop in the rookery. Oddly enough, though, the eggs when newly laid can survive a substantial amount of cooling in freezing water.

In some years a change in the direction of wind-drifted snow or unusually large melt streams can keep all the eggs from hatching in a rookery or portion of it. When this happens, people begin to worry about the birds' survival. We should remember, however, that the Adelie lives about 12 years on the average, and if a female starts breeding at three or four years and a male at four or five, quite likely they will raise at least two chicks during a lifetime. If each breeding pair reared two chicks each year, the Antarctic would soon be overrun with penguins.

High mortality of young is understandable in such a harsh climate. Only 10 of 100 Adelies live to the age of three to return to the breeding ground where they were hatched. Precise marking combined with consistent observation over a number of years has given us much insight into the population dynamics of the species.

The feeding trips to sea by male and female are closely coordinated during the breeding cycle. The female incubates the eggs for the second two-week period while the male feeds; then he returns to the nest for a shorter tour of duty—a week or less. By late December the ice foot has broken away. Males now gather on shore, each reluctant to take the first plunge into the churning water and ice, where a seal may be lurking. This will be a short trip. New demands will be made on both mates when the chicks hatch.

A certain restlessness prevails in a penguin colony at hatching time. On a windless day you can even hear the peeping of penguin chicks from inside the eggs. And, especially on a warm day, the incubating parent may stand up and call out to the noisy eggs or give the ecstatic display, perhaps expressing the nervous tension of imminent parenthood.

Although the two eggs of a clutch are generally laid two or three days apart, they hatch within 24 hours of each other. This is difficult to explain, because the adults constantly incubate both eggs at a temperature of about 100° Fahrenheit. Perhaps a clue comes from studies of such birds as pheasant or quail where it has been found that the eggs "talk" to each other. Peeping sounds between eggs can actually speed up the hatching efforts of chicks that are a little behind, thus coordinating hatching time.

Nest reliefs occur much more frequently when the downy gray chicks emerge. Parents alternate in guarding the hatchlings and gathering their food, regurgitated in small amounts throughout the day. During the chick-rearing period the parents make 35 to 40 journeys, each involving at least 24 hours at sea pursuing swarms of edible marine life. The bird swims out at about 10 miles an hour, covers 20 to 200 miles, and may devour over 5,000 organisms.

Back at the rookery you see only very fat penguin parents lumbering up the hill, bringing home the krill. Often they pause and

While summer breezes blow snow on its back, an incubating bird bristles with resentment at the photographer's intrusion. Its alternate stare threat display—bill down, eyes rolled, neck arched—may precede an attack with bill and flippers. Midsummer days go only a few degrees above the freezing mark. A featherless incubation patch on the sitting adult helps keep the eggs at a constant temperature.

ROBERT W. MADDEN

preen to oil their feathers, ensuring dry insulation for the next trip into the icy waters. The chicks are seemingly insatiable. A four-pound chick can eat an equal weight of food within 15 minutes. For the first three weeks either the male or female parent constantly warms and guards the chicks. This guard stage is crucial for forming bonds of recognition between parents and their young. Here the voices of parents (and probably also of their chicks) again play a useful role in penguin society. The ability of parents to distinguish their own young is paramount during the next part of the cycle, which is the crèche stage.

When the parents first leave the chicks alone and both go to collect food, the youngsters gather in groups called crèches, consisting of a few to several dozen downy Adelies. The compactness of the group varies. On a bright warm day they spread out; on cold, windy days or when danger threatens they huddle close together.

Nest becomes nursery after an
average 35 days' incubation.
The proud parent on duty inspects
a new arrival (opposite, upper).
Downy chicks weigh in at less
than three ounces. Hatchlings,
fed on semi-digested seafood,
grow so fast that at two weeks
an adult can barely cover them.

A raucous relief ceremony marks
the changing of the nest guard
(above). Chicks chime in,
acquiring traditions that help
the species keep to the closely
timed breeding schedule dictated
by the polar environment.
Family members recognize each
other by voice—vital in the
crèche stage, when the generations
separate. Both parents are
busy getting food; chicks gather
in crèches—peer groups.
Adults flanking the small crèche
(center) likely are bachelors
claiming sites for next year.

To the human eye all chicks look alike. It would appear that any adult returning from the sea feeds any chick. The adult, bulging with food, comes up to the edge of the breeding colony or to the site of the vacated nest and gives the loud mutual call, a hoarse *aaaah aaaah aaaah,* and chicks come running. I heard the calls repeated so often during the rearing of chicks that I thought of a possible reason why penguins are so noisy: that the voice is the means of identification between parents and young.

By accurately marking entire Adelie families (with bands, colored tapes, string, or paint) and spending many hours recording their individual voices on tape, I prepared to test the hypothesis. In quiet Colony IV at Wilkes Station on a warm afternoon I played the recorded calls of a particular parent through a loudspeaker. The response of the chicks belonging to that adult was spectacular. They suddenly jumped up from the crèche of lounging chicks and ran to the location of the nest they had been hatched in. I can only describe their appearance as having an "expectant air." In nine of ten playbacks, using seven different recordings, the offspring of the adult whose voice was played reacted similarly, while no other crèche chicks so much as changed position. The tenth playback, which brought no positive reaction, was of an adult whose chick had recently been fed. When I accidentally played one tape backward, no chicks responded except to lift their heads.

Sometimes other hungry chicks in the crèche will chase a returning adult along with its own chicks. Many observations, mine among them, show that a parent rarely feeds a chick other than its own.

Intricate feeding chases aid the sorting-out process. At the beginning of the crèche stage the raucous yelling of food-laden adults and the begging peeps of chicks are followed by short circle chases around the old nest site. As the chicks get older, the chases get longer. The parent pauses now and then to peck away strange chicks. The legitimate offspring continue running and are fed during pauses. If there are two chicks, the fastest gets the mostest. But as the fast one gets full, a low-hanging stomach slows it down. The thinner chick then has its chance and is fed a fair share. Raising two chicks can be a particularly arduous task!

In a good year the average breeding pair raise only one chick to self-sufficiency. Many pairs lose both eggs or both chicks. Unusually heavy ice along continental shores can make the trek to sea for food too long. The loss of a parent at sea usually means a chick will die; one parent can hardly get enough food for two.

Extreme weather may also kill chicks. Devastating blizzards can literally blow chicks into the sea with their down iced into a white coat. At Wilkes, winds of 115 miles an hour were common and we sometimes had gusts in excess of 170 miles an hour. At other times, hot summer days of 40° Fahrenheit may put downy Adelie chicks into fatal heat prostration.

Roly-poly chicks show patches of white chin feathers that mark the start of molt. Final food chases, when chicks are seven or eight weeks old, lead them to the sea's edge. Nearly grown now, fully feathered, and deserted by parents, they must soon take the cold plunge and fend for themselves.

DRAWINGS BY GEORGE FOUNDS

The menace of the south polar skua is yet another story. This fascinating gull-like bird of hooked beak and slashing talons also breeds in Antarctica in summer and has its own young to feed. It is both predator and scavenger. One study shows that 30 percent of its feeding during the austral summer involves penguins. The rest of its food comes from organisms living in the sea. Skua attacks on crèche chicks can range from the seizure of starving ones to an organized assault on a healthy chick.

At Wilkes Station the skuas not only maintained strict breeding territories but also feeding territories. The late Carl Eklund and I learned the hard way about skua defense of territory. Carl's studies called for marking all the skuas at Wilkes with a colored plastic band on one leg and a metal band on the other. When a human being invades the breeding territory the skua responds by dive-bombing. The bird calls loudly, flies high and comes down at a speed which cannot be less than 40 knots. If you don't have a hat on, you get knots on the head from the skua's feet. The bands on their legs accentuate the pain.

I became a particular target of such dive-bombing when I was trying to maneuver a motorcycle across the snow. It was a strange experience. One highly territorial male skua would repeatedly give me a hard time as I approached the rookery on my cycle or whenever I invaded his real estate to make observations in Colony VII.

Curious mealtime ritual, the feeding chase, gives crèche chicks a workout. The voice of a food-laden parent calls the downy youngsters to the nest site (opposite). The adult then turns and runs away. Chicks pursue and are fed beak-to-beak during pauses. The circuitous runs around the rookery gradually lengthen as the young birds grow stronger. Ravenous, at five weeks they can consume their own weight in food—about four pounds—at a meal.

Fierce south polar skua swoops down on a lone Adelie chick. Though vulnerable away from the crèche, the intended victim (above) stands its ground, threatens back—and escapes. Adult penguins ignore the plight of a less fortunate chick (center and opposite). Caught by its own beak, it is borne away by four-foot wings to the skua's nest. Fearless even of man, this gull-like raider usually kills a penguin chick by pecking it on the head. Its toll includes a high percentage of the weak and those chicks that tend to wander from the group.

DIETLAND MÜLLER-SCHWARZE
ABOVE: RICHARD L. PENNEY

Inevitably, as I lifted my binoculars to read a penguin band number, I would get clobbered on the head. After portions of three consecutive seasons of this form of behavior from the same skua, it took all my patience to keep from using the shotgun in the name of legitimate biological investigation. Finally, I took Carl's advice and added a feather to my hat. The skua seemed content to strike that with his legs, leaving my head somewhat relieved.

For penguin chicks, such harassment is not only severe but often fatal. A hungry pair of skuas on their feeding territory at Wilkes could easily isolate, kill, and eat a perfectly healthy crèche-stage chick. Naturally, a sick chick proved easier prey. With respect to skua predation on penguins, I strongly suspect that different behavior exists depending on local conditions. Unfortunately, many naive people, out of personal preference for the penguin, look upon the skua as evil. Skuas have been harassed severely at several locations in Antarctica. This human interference affects the normal balance of the predator-prey system.

The field biologist makes no moral choice between "good" penguins and "bad" skuas, leaving the two populations of birds to do their own thing. I must confess to a certain thrill, however, when an Adelie chick under attack puts up a bold front, returns threat for

threat, and manages to get back to the crèche where there is safety
in numbers—and the incidental protection added by the presence
of adult penguins.

William Sladen first tested the long-held belief that specific adult
penguins were left behind at the crèche to guard the chicks from
skua predation. Dr. Sladen marked chicks and adults and found that
there was not a guard system as such. Parents recognize and feed
only their own young; the other adults about the crèche are there
primarily to claim nesting territories for the next season.

Further studies, however, have shed more light on this behavior
pattern. At Wilkes and Cape Crozier I found that adults that had
lost eggs or chicks early in the season and were reclaiming territory
did in fact produce some guarding effect. They would rush out from
their nest sites to charge a skua and scare it off. I don't think we can
say they were purposefully guarding chicks, but their defense of
territory did add to the chicks' protection. We may have here one
of the fringe benefits of colonial breeding.

As the chicks mature and gain their first-year juvenile plumage,
the feeding chases extend to the beaches. Here, away from the pro-
prietary atmosphere of the colony area, the Adelies again become
highly gregarious birds. The chicks congregate, flap their flippers,

gaze out to sea, rush down to the water's edge, then chicken out—or "penguin out." Finally, of course, they must go to sea. There are no more free meals on the way!

The chicks leave alone or in groups along with adults. The chicks' first try at swimming in icy seas is not very impressive. The adults scoot off at about ten miles an hour—swimming underwater, porpoising for air—to the distant floating belt of sea ice surrounding the continent. The chicks bob along on the surface, making clumsy attempts to porpoise, becoming easy victims for the leopard seals. From one cause or another, the majority of the chicks will die before they reach breeding age. Nevertheless, the Adelie is an extremely successful species. I wouldn't care to guess how many millions of them are in existence.

Dispersal and departure of the chicks takes a surprisingly short time. The field biologist actually feels very let down when the pandemonium and super-activity ends and the rookery is quiet again. Most of the chicks that survive the long winter trip will remain on the pack ice for the next two or three years. Many will eventually return to their natal rookery to breed and become the involuntary subjects of penguinologists' experimentation and study.

We still lack complete knowledge of movements and activities away from the rookery. We do know that penguins, like many other species of birds, use a "biological clock" and the azimuth position of the sun to tell directions. This undoubtedly helps them to find their way back to their nests year after year.

The final event at a rookery is a passive one—molt. After their hard work of rearing chicks, adults spend a couple of weeks at sea, overeating. Then you find the fattest penguins of the season. The reason is simple: For 21 days they must go without swimming, or food, while they drop the old feathers and grow new insulation for the impending winter. During molt they lose up to 50 percent of their body weight just sitting on a wind-protected snow or rock slope, unable to migrate without mature, well-oiled plumage. This is a delicate part of penguin life history.

The temperature is dropping rapidly. Winter winds are beginning to blow. By the first of April the cycle is completed. The rookery becomes as lonely as the period in mid-October when the leopard seals first prowl the beaches for penguin and man.

At summer's end, the Adelie penguins depart to slightly warmer seas off Antarctica. While the continent lies icebound, they winter on pack ice, whose northern rim reaches to about 60 degrees south latitude, feeding in the sustaining sea. They migrate in mixed social groups of males and females. Families separate, the task of bringing forth a new generation now ended.

ROBERT W. MADDEN

THE MARVEL OF LEARNING

To the delight of his audience at a midwestern fair, a young pig picks up a large wooden coin with his mouth and drops it into an outsize piggy bank. With the crisp efficiency of a teller he banks four more. No ham actor, he ignores his public's applause and trots to a food dispenser for his real reward.

No pig in his sty ever put coins in a piggy bank. Nor did this one until rewarded for it by Keller and Marian Breland, psychologists who began training animals commercially in the 1940's. Through experience, the pig changed his behavior—a process we call learning.

Most animals, from insect to man, learn. And they do it essentially on their own. It takes two to mate, two to fight, two to communicate; only individuals learn. Other animals may help or distract, but they do not determine how the learner learns. Subtle chemical changes, a faint pulse of electricity from one cell to another—by such minute transactions does the brain perform its mysteries. Until scientists can measure exactly what happens in a brain when an animal learns, they can only observe the effects.

One scientist finds a niche in the wild to study a species' natural history. Another experiments in a laboratory, gauging behavioral patterns by hundredths of a second or millionths of a volt. The first speaks with the ring of ecological validity, the second with the authority of precision and control. Together, as we see in the following pages, the two approaches build man's knowledge of learning as neither could do alone.

Sometimes the findings are surprising. The thrifty pig performs flawlessly for weeks, even months. But then, instead of banking the coins, he begins to drop them, root at them, toss them in the air. Such ad libs cost him his reward, and his performance worsens. Eventually he beds down at night a hungry pig, having banked too little to earn a day's pay.

Pigs were not the Brelands' only wayward pupils. A raccoon started to learn a similar task but became so engrossed in fondling the coins that it never finished. A hen learned to bat a ball across a miniature baseball diamond by pulling on a rubber band, but the game was called when she began to field her own hits and peck strenuously at the ball. A cow would not learn to deliberately kick a bucket, a goat would not bleat for food, and a pigeon would not pick up a loop of string from the floor.

Each in its own way showed that we cannot teach all things to all creatures. The pig, the raccoon, and the hen drifted into the behavior they show toward food in their natural habitats. Evolution has not programmed the brains of the cow, the goat, and the pigeon for acts another animal might learn readily. Yet in real life each learns what it must.

Without the ability to learn, animals would lose the struggle for survival. For no two events unfold alike. Different hungers, different perils, different opportunities to gain a tiny edge present themselves every day. Truly, we all live and learn.

A rat masters a maze, a chimp learns a symbol language, and from each men sift secrets of the brain. 357
PAINTING BY RICHARD SCHLECHT

LEARNING
HOW ANIMALS LEARN

Arthur J. Riopelle

A monkey glowers from a cramped cage as it paces to and fro. By accident it steps on a pedal at one side of its prison—and a door on the other side flies open. Freedom! The monkey scampers out with obvious satisfaction. A researcher shuts it in the cage again. This time the monkey gravitates toward the pedal side; in less time than before it hits the pedal and escapes. Again and again it frees itself, each time more quickly than before, until the researcher can barely close the door before his "student" escapes.

The monkey's task could just as well be to pull a lanyard, do a flip, or push buttons in proper sequence. It might win food, a chance to mate, or playtime with a toy. When E. L. Thorndike devised the basic "problem box" procedure in the 1890's, it seemed that almost any response could be learned if it were rewarded. When dogs, cats, and even fish mastered tasks set for them, he concluded that linking an act with a reward was the basic form of learning, and that improvement occurred gradually.

Meanwhile, the Russian physiologist I. P. Pavlov probed a similar kind of learning called classical conditioning. He studied how dogs salivate when they eat, and then found they would drool at the blink of a light or the tick of a metronome if such neutral stimuli appeared often enough with food. And he too found that he could condition several kinds of responses in a variety of animals—chickens, rats, even human beings.

We have learned much about learning since then. We know now that we cannot link just any response to any stimulus. For every creature interacts with only a fraction of its environment. Be its niche broad or narrow, an animal shows behavioral limitations in the laboratory corresponding to those in the wild. If the laboratory conditions overstep the bounds of the animal's niche, the experimenter is unlikely to get the expected response.

Almost any monkey can learn to press a red button and not a blue one for food. In fact, it is by such discrimination-learning experiments that we have learned that most monkeys see colors about as well as humans do. The best-known exceptions are the night monkeys of South America, whose visual system is adapted for seeing in dim light rather than for seeing colors. In such light we humans can't see colors either. Color discrimination, then, is an ability not needed in the night monkeys' niche, and no amount of training will teach them to discriminate red from another hue.

Or think of a duckling emerging from an egg. The chances that a coordinated activity like walking, involving a hundred or so muscle groups, would spontaneously result from random activity are infinitesimal. Yet the duckling walks. Evolution has built in an unlearned behavioral sequence that helps the species survive. But we know of no way to teach a robin or jay to walk or fly at hatching. Until they mature sufficiently, the mechanisms for locomotion are inadequate; such birds survive through dependence on parents.

Each animal's nervous system and behavioral capacity mature according to its genetic timetable, and learning potential is closely linked to the process. In general, some aspects

Is it a cliff? The eyes say yes
but the paws say no—they can feel
the glass that would carry the
kitten across. The eyes have it:
The kitten avoids a "visual cliff"
almost from the moment it begins
to walk. But if reared in darkness
it must learn in the light to let
its eyes guide its feet. Many
species—even human toddlers—
have been tested on visual cliffs.
All of them except the stoutly
armored aquatic turtles heeded
most visual warnings when old
enough to move about.

Is it a hawk? Towed overhead
blunt end first, a cutout shaped
as below resembles one. In a test
it sent wild ducklings scurrying
for cover. Reversed, it suggested
a long-necked goose; only half
the ducklings displayed alarm.

Many kinds of young fowl show
fear toward cutouts of widely
varying shapes, sizes, and speeds
—and even toward a falling leaf.
Some, such as young turkeys,
habituate to shapes rapidly but
will flee from ones that are rare
in their environment. By such
experiments scientists seek
to reproduce and understand
meaningful natural experiences.
MARTIN ROGERS

Goose ⟵   ⟶ Hawk

of the nervous system trail others in maturing. A robin hatchling cannot walk, but it can breathe, chirp, purge wastes—and, as harried parents discover, it can eat.

Just as humans creep before they stand and stand before they walk, other species pass through stages of locomotion timed largely by the maturation of their nervous systems, with learning only refining the behavior. In a convincing experiment that remains a model of its kind, Leonard Carmichael kept one group of young salamanders immobilized with an anesthetic until their freely moving counterparts in a second group began to swim. The first group swam too when released from the anesthetic, even without passing through the stages that normally precede swimming.

Touch a newborn human infant on the side of its face and it will turn its head. When Dr. Carmichael, in a study of fetal guinea pigs, touched an unborn animal on the lip, it brushed away the stimulus with a paw. When the stimulus was applied to a shoulder, the guinea pig's paw again brushed it away. But among the lower invertebrates an animal's first responses to such stimulation are large movements involving much of the body; only later does it move specific parts.

Experimenters have further probed maturation and learning by restricting experience or making it unusually rich. Chicks raised in isolation avoid other chicks when placed among them, and fail to react to a hen's food calls. Chimps reared in darkness need at least two weeks in the light before they learn to heed visual warnings of an electric shock. But rats raised in the dark get along very well in the light, suggesting that light stimulation is more important to the diurnal chimp than to the nocturnal rat.

Animals given enriched experiences tend to be calm and outgoing, always ready for greater contact with their surroundings. Usually, the earlier the enrichment the greater the effect. In fact, the brains of rats reared in impoverished surroundings actually differ

Class is in session as B. F. Skinner places a pigeon in a "Skinner box," a version of the "problem box." In such a box another pigeon shows it can learn to discriminate between similarity and difference. In three peepholes, combinations of 12 patterns appear. The bird's job is to peck two that match. When it does, it hits the jackpot; a buzzer sounds and the bird collects a reward of grain. Some birds peck winning combinations 40 consecutive times or more.

Dean of behaviorists, Dr. Skinner in four decades of work with animals, and lately humans, has defined a provocative thesis: Man acts not by his own free will but by conditioning to the environment.

MELVILLE BELL GROSVENOR AND (OPPOSITE) VICTOR R. BOSWELL, BOTH NATIONAL GEOGRAPHIC STAFF. LEFT: THOMAS NEBBIA

in their chemical composition from those of rats brought up in enriched environments.

After decades of experimentation with simple conditioning, we know that most animals learn conditioned responses at about the same speed as did Pavlov's dogs. Thus what we learn even today about conditioning in the lab, though observed directly in only a few species, can often be applied to the many species in the field. But such factors as environment and experience also determine what, and how fast, a given animal can learn.

Among humans, innate behavior patterns are more prominent in infants, while learned behavior governs adults. And theoretically, the more complex its nervous system, the less a creature tends to depend on inborn actions and the more it relies on learning. This theory led to studies of problem-solving intelligence not only among the manlike primates but among other animals as well.

Some scientists cautioned that Thorndike's problem box never revealed insight—the flash of intelligence which solves a problem in an instant—because the box kept the animal from seeing the relationships among the elements of the problem. Hence they devised the detour method: Block the direct path to a reward but leave open a roundabout way. The animal displays insight when, after a pause filled with unproductive activity, it suddenly takes the detour and collects the prize. A chimp will scoot around a wire fence to get a tidbit, but a chicken similarly thwarted will wander about and find the food only by accident. One researcher hung a banana from a ceiling to see if a chimp, already used to boxes, would think of stacking some that were nearby to reach it; instead, the chimp took his hand, led him under the banana, and climbed up *him* to the prize!

Another type of test requires an animal to react to something that is absent—to remember. A caged raccoon faces three doorways and is trained to get food through whichever one lights up. Later, when we light one doorway for a moment and release the raccoon 25 seconds after the light goes out, it walks to the proper doorway.

Primates show a special ability to benefit from experience in problem solving. Monkeys improve from one problem to the next; given enough experience, they will tackle a new problem and solve it virtually without a mistake. Rats, raccoons, and cats also improve with practice, but considerably less than do rhesus monkeys. A rhesus adept at problem solving can watch another perform a single trial on a problem and then get it correct even if the other got it wrong. An adept rhesus will also learn faster how to avoid a shock when a light appears if it first watches another rhesus learn how.

What happens if we train an animal—say, a rat—to press a lever for food, and then start withholding the reward? You can answer this from your own experience. How many times have you opened the same drawer again and again to look for something that's missing? You and the rat keep making the same response, tapering off only gradually until, in the jargon of the laboratory, the response extinguishes. Thus a learned response may persist even without a reward—a fact painfully well known to every gambler.

Monkeys new to discrimination-learning may take 20 trials to learn that an ashtray and not a cup hides a raisin. Curiously, they learn faster if there are several cups rather than one. The ashtray stands out from the crowd; it becomes a more salient stimulus.

Some stimuli are naturally attractive. Light-catching hair, a freckle, a cuff button—each a salient spot—can evoke grooming behavior in chimps, (Continued on page 372)

LESSONS FROM APES

Snowflake, the only albino gorilla ever known to science, studies a blue-eyed reproduction of himself—while researchers study him. On a visit to the Barcelona zoo, Dr. Riopelle gave the doll to the young ape and watched his reaction. Soon Snowflake grabbed it, embraced it, and gently carried it about. When Dr. Riopelle took the doll apart, Snowflake's interest focused on the head; the torso got slight attention, the arms and legs even less. Though wear and tear took a toll, the doll's appeal persisted for several years.

Such spontaneous attachments play a key role in determining the direction and speed of learning in the wild and in the laboratory. Both are habitats of scientists observing primates, whose brainpower makes them ideal for studies into the nature of learning.

363

Brothers under the skin

Is Snowflake different from other gorillas? With support from the National Geographic Society, Dr. Riopelle returned to Spain periodically to seek answers through an in-depth study of the unique albino, captured in African lowlands in 1966. Snowflake apparently is like other normal gorillas except for his coloration and the slightly impaired daylight vision often found in albinos. And, like other gorillas, he maintains his own individuality.

Clapping wildly, Snowflake expresses elation much as a human would. His increasing weight and activity have outreached the patience and strength of researchers in their sessions with him.

In his younger years, Snowflake contributed his bit to our knowledge of albinism's effects. His impaired day vision, for example, may have robbed him of youthful confidence; his cagemate Muni dominated most of their activities and sometimes outshone him in Dr. Riopelle's tests. Given a mirror, Snowflake ran from his first encounter with the glass, then later beat at his own image as if testing its reality. But Muni strode straight to the mirror. Fascinated by his reflection, and seeming to realize that the image was not a real gorilla, he used the glass as a tool to inspect parts of his body that he could not ordinarily see.

One day a zoo technician dyed her hair red; both gorillas reacted to the change. Learning of the incident, Dr. Riopelle devised a novel test: He walked into the cage, a familiar figure suddenly transformed by a garish blond wig. Snowflake eyed the wig, reached out, drew back; he took many minutes to get used to the change in his old friend. Muni yanked the wig off in a trice.

Snowflake now shares his cage with Endengui, a female donated by the Society. Science hopes they may start a line which could produce more white gorillas to aid in studies of albinism, an inherited condition found among animals from mice to elephants and whales.

MICHAEL KUH

In a barren cage a young rhesus monkey curls inward upon itself. Reared in isolation, bereft of toys or other items of interest, fed well but not by a mother, it shrinks from a world it will never accept.

Freed from isolation and caged in a simian playland, monkeys from impoverished environments withdraw socially. Sometimes they cling together (right) while normal youngsters romp around them. Their actions are stilted and repetitive; they do not groom properly or "read" the actions of others. Many will never mate; females that do mate make hopeless mothers, even spurning their young.

As with monkeys, so with men: Much of life's pattern is shaped in childhood. But what factors mold a child as it matures? Scientists cannot rear human infants in laboratories like this one at the University of Wisconsin. And so they turn to our primate relatives. The rhesus monkey, a species biologically similar to man, helps psychologists identify those formative factors.

Deprived of mothers but given companionship of their peers, rhesus monkeys outgrow a slow start and eventually develop normal social behavior. But isolated from all social peers from birth on, a young monkey grows up seriously incapacitated. Some may be helped by what Wisconsin's Harry Harlow calls therapist monkeys. Younger, smaller, and without their patients' hangups, they may in time begin to draw out the disturbed animals.

Just to see something happening is a powerful reward for a monkey. Placed in a plain cell, it will work to open a window for a glimpse of another monkey, an electric train, or even a busy lab.

Toys, blankets, and soap opera enrich the days of a young chimp at Yerkes Regional Primate Research Center in Georgia. Reared in a rich environment but without companions, a chimp may become a slow student, never fully learning how to learn—much like a child given inadequate stimulation in infancy. Such studies, which help in planning preschool programs, remind us that if certain kinds of learning do not occur early in life, they may never occur at all.

Growing up in the lab's many worlds

MARTIN ROGERS

367

Re-inventing the ladder

At first the poles were only playthings placed in a pen by Emil Menzel at the Delta Primate Center in Louisiana. But as a group of eight chimps romped with the newfound toys, the apes got the feel of each one. They would stand a pole on end and scoot up it or vault into the air, tumbling to the grass just for the fun of it.

Several years later, one chimp climbed a pole to get something off a shelf. The toy had become a tool, and soon all knew how to use it. The procedure was simple: Fetch a pole, lean it against the obstacle, adjust the angle, and climb. The effects were sometimes calamitous. No longer would electrified wires around trees and at the base of an observation house keep the chimps from climbing up to peer at observers or to reach branches and strip them of tasty leaves. One night the truant apes broke into the house; when technicians chased them out one window, they promptly climbed a pole to another. Two weeks later all eight were discovered in the trees, with a telltale pole leaning against a trunk above the encircling wires.

Insight and experience produced the invention; imitation spread it throughout the group. But the chimps' grapevine can spread news by subtler means. Dr. Menzel put a dead snake in the pen and let in one chimp to discover it. Next the chimp was returned to the group, the snake was removed, and the group let in. Several grabbed sticks and attacked the spot where their comrade had been frightened.

Apt mimics of man, chimps can learn to play games, drive cars, and paint pictures that impress art critics. In centers like this it is man who learns from the chimpanzee, perhaps our greatest ally in the quest for a better understanding of ourselves.

Chatting with a chimp

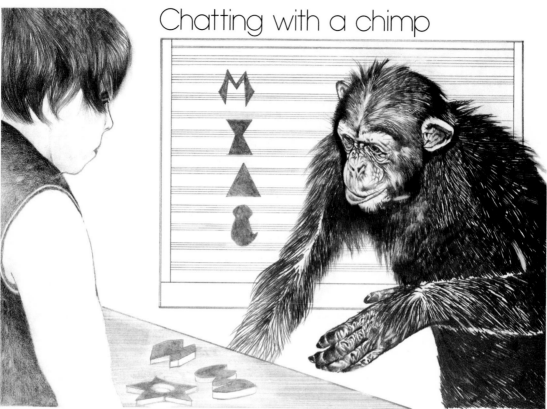

DRAWING BY GEORGE FOUNDS

APPLE NAME OF

BANANA NOT NAME OF

SARAH

JAM

BREAD

TAKE

NO

SARAH

HONEY

CRACKER

TAKE

Sarah the chimpanzee turns back to her tabletop vocabulary after putting four of its words on a magnetized board beside her. The words—metal-backed plastic symbols—read top-to-bottom, Chinese fashion, because Sarah happened to begin that way when she learned to write. They tell her trainer plainly: MARY GIVE APPLE SARAH. And of course trainer Mary hands Sarah a slice of the fruit.

Sarah's sentence is short. But its import is startlingly long: An intelligent primate can communicate with man in a human way, using symbols—man's own words are symbols—arranged in sentences.

David Premack, a psychologist then at the University of California at Santa Barbara, shaped Sarah's literary education. He began by analyzing our language to compile basic features he called "exemplars." Then he designed ways to teach them to Sarah. Some she learned were: words, sentences, class concepts (shape, size, color), *if/then* relation (if X happens, then Y results), and metalanguage (the use of language to teach language).

Sarah's feeding routine made a starting point for her training. At first the instructor offered a slice of banana and looked on affectionately as Sarah ate. This established a social exchange and a favorable learning situation. Then a plastic square was put in the banana's place and the slice was moved out of Sarah's reach. In order to have her banana, she had to learn to stick the square onto the magnetized board. Thus ☐ came to mean *banana*.

The social transaction of giving consists of a donor, an action, an object, and a receiver. Sarah had named the object. Next she was taught to identify the donor, Mary, with a plastic ⋈. And she could

SARAH

INSERT

BANANA

PAIL

APPLE

DISH

SARAH

BANANA

TAKE

IF/THEN

MARY

NO

CHOCOLATE

GIVE

SARAH

write 🐍 ☐, a two-word phrase. In time she learned the remaining elements and received the banana only on writing MARY GIVE BANANA SARAH. Later, more complex sentences instructed her to munch on jam and bread and ignore a honey cracker (opposite page).

Dr. Premack used the concepts *same* and *different* to acquaint Sarah with interrogatives. Sarah's ability to pair like objects first was tested. Given two oranges and an apple, she paired the oranges. Then she was taught to place the plastic word SAME 👀 between like objects and DIFFERENT 👀 between unlike ones. Now a ? symbol was used between the objects so that Sarah in effect was asked: "What is the relationship between orange and orange? Is orange the same as orange? Is orange the same as apple?" Sarah learned to replace the ? symbol with SAME or DIFFERENT, and with YES or NO. Thus training gave her words to express an ability she already had.

Teaching new words via a social transaction went slowly; metalanguage sped the instruction. Putting a symbol between an actual object and her word for it taught Sarah such concepts as APPLE NAME OF a real apple and BANANA NOT NAME OF a real apple (opposite page). Thus the language phrase *name of* was used to teach a language word. Sarah quickly put the system to use; she adopted FIG and CRACKER JACK for foods she liked but had not previously named.

Class concepts—color, shape, size—proved no problem. Asked RED ? APPLE, Sarah learned to reply RED COLOR OF APPLE. Introduced to the verb *to be,* she could specify RED IS COLOR, RED IS NOT SHAPE. Mastering pluralization, she could write RED YELLOW IS PLURAL COLOR (red and yellow are colors).

The *if/then* relation was more difficult. To teach it, Sarah was rewarded for doing one thing but not for doing another. Then words went into the situation. Ultimately, she could work out 17-word instructions offering a chocolate if she correctly chose between an apple and a banana. (Half the sentence read SARAH APPLE TAKE IF/THEN MARY CHOCOLATE GIVE SARAH; the rest appears at left.)

Compound sentences answered the important question of whether an animal could understand sentence organization. To interpret correctly "Put the banana in the pail and the apple in the dish, Sarah" (upper left), she must know that *banana* and *pail* go together, not *pail* and *apple*. And she must understand that it is Sarah who has to perform the action on both fruits, even though the action word ☆ appears only with *banana*. Sarah's performance showed that a sentence to her had order and was not a haphazard string of words.

What do these plastic pieces mean to Sarah? Consider first what *apple* means to us. It signifies a sweet, round, red fruit with a stem and seeds—not a five-letter word with two p's. Sarah sees a real apple much as we do. And she describes △, her word for apple, as round, red, and with a stem! Sarah's pieces of plastic, her words, are as much symbols to her as our words are symbols to us.

Sarah's comprehension of a number of language exemplars proves her success in the main task of her education: to show that she can understand what is being said. Now she has graduated to a typewriter with symbols on its keys. She and her instructor can chat via keyboards that produce Sarah's symbols on a television screen. She may have some interesting things to write in days to come.

At age 9, after two years' training, Sarah had learned 120 words. Here is a sampler:

prefer
cut
cook
eat
drink
insert
smoke
is
pl (plural)

orange
peach
pear
fig
apricot
grape
date
milk
sugar
cookie
caramel
Cracker Jack
peanut
peanut butter

key
table
dress
shoe
cup
crayon
paintbrush
paper
flashlight
broom
dustpan
basin
soap
comb
toothbrush
toothpaste

yellow
blue
green
red
orange
brown

good
bad
big
little
round
square
triangle
one
several
all
none
much
yes
no
name of
size of
shape of
color of
same
different
on
in front of

even those born in the laboratory. A shadow which suddenly looms larger causes many animals to go into a defensive posture. In many creatures such "triggers" are so specific that special neural mechanisms exist to handle them. Researchers have inserted tiny electrodes into single nerve cells in cat and monkey brains to study the role of these cells in perception. They have found that certain brain cells react only when the animal sees lines at a particular slant; other cells respond only to edges of light, still others only to movement in a specific direction.

Some brain cells react to high-pitched sounds, others to low, still others to sounds which might be communicative. In monkey brains, specific cells react strongly to the first in a series of meaningless clicks but only weakly to the last, whereas they respond continuously to monkey calls. A familiar voice repeating a command also provokes a persistently strong response, but a strange voice doesn't.

Emil Menzel in a field study threw a plastic orange to a troop of rhesus monkeys. As they chorused the typical food bark, a dominant female raced to the orange. When she realized something about it was unusual, she jerked her hand away, peered at the object, and barked the avoidance call. Instantly the others quieted down and avoided the bogus orange. Did she "speak" to them? Probably not, at least not in the human sense; the others simply knew from lifelong experience that when one of their number calls out in alarm, avoidance is the prudent response.

In a one-acre outdoor enclosure at the Delta Regional Primate Research Center, Dr. Menzel carried a young chimp to some fruit hidden in tall grass, then returned him to his group in an adjacent cage. Within seconds after release, the chimp "in the know" scampered toward the food with the whole group at his heels, every follower intently watching the actions of the leader. When two leaders were shown different food caches, they seemed to pool their information, for they normally went first to the better cache. If the leaders split, the one going to the better supply usually attracted more followers. Most likely this communication was simply the difference in the intensity of the leaders' recruiting methods —running, whimpering, gesturing, tugging—coupled with the ability of animals to interpret each other's behavior.

Compared with human speech, facial expressions, and social etiquette, communication among animals may seem rather a simple affair. But the great apes may have a capacity for communication far beyond what we observe in the field or zoo. Speaking is beyond them; some scientists believe the larynx is too high in the throat to permit the necessary vocalizations—as perhaps it was for Neanderthal man and is for the infant *Homo sapiens* today. Viki, a chimp raised by a research team, never got far beyond "mama," "papa," and "cup." But Sarah and Washoe, the chimps that "talk" with plastic symbols and sign language respectively, prove that chimps can communicate by human means.

The enormous skills and intellectual prowess of apes and monkeys should give man considerable pride in sharing ancestry with them. In laboratory tests of learning, problem solving, and communication they far outstrip the requirements of everyday living. Much of their untapped brainpower is not particularly needed for survival in forest or savanna. Is it any surprise that many thoughtful people wonder whether man, with even more of this excess potential, may be too smart for his own good?

Hands do the talking for Washoe,
a chimpanzee reared by Allen and
Beatrice Gardner, psychologists
at the University of Nevada.
The Gardners reasoned that, since
chimps are adept at gestures,
one might be taught to
communicate using the American
Sign Language; its hundreds of
gestures convey not letters
but words and ideas.

Born in the wild, Washoe began
her education when about a
year old. Teams of trainers fed,
groomed, and played with her
daily, using no language but ASL.
Within a year she could link signs
in simple expressions like "gimme
drink," to which she soon learned
to add "sweet" when she wanted
soda pop. After four years she
had a vocabulary of 130 signs—
and a few phrases of her own
invention. Spurning "cold box,"
she asks trainers to "open food
drink," her name for refrigerator.

Washoe starts conversations,
answers questions, and apologizes
for mischief or accidents. She
even talks to herself. Unaware
of being watched one day,
she signed "Washoe climb tree,"
then climbed one. Chatting with
a trainer, she says (from top)
"flower," "baby," and "hear"—
with gestures that come close
to these in the ASL lexicon:

Deaf persons read her easily,
despite her slight chimp "accent."

As other chimps learn ASL, their
mentors ask an intriguing question:
Will chimps in a group talk in signs?

T. J. KAMINSKI

373

LEARNING TO LIVE

William A. Mason

Most anyone who has lived on close terms with animals, whether dog owner and his household pet or spangled trainer and his circus cat, has been impressed with how much they learn—about us. The small details of the daily round, our fleeting moods and subtle gestures can be as meaningful to them as the command to sit, to fetch, to stay, or to jump that we pride ourselves on having taught them.

The dog is pre-eminently a teachable animal. I suppose we humans deserve some credit for this characteristic. After all, we have selected and bred our dogs through countless generations for traits that we find useful or pleasing. But surely a great part of the credit for the dog's adaptability, its remarkable talent for learning to live with others, belongs not to man but to its ancestor, the wolf. This pack-living hunter has a life-style that places a premium on the ability to learn from, and about, and with others.

In man's long history of living and working with animals, he has not always succeeded in training them to his purposes. But even his failures have taught him something about how animals learn to live. A young pig quits banking a coin with his mouth in a fairgrounds act and begins rooting at it; a trained chicken pecks at a miniature baseball instead of using it to win a reward. It is as if older patterns, despite the training, eventually surface to interfere with performance of the learned act. The mark of the wise trainer is that he tries to avoid such interference by fitting his training techniques into the animal's existing behavioral programs. This often involves entering into a close social partnership with the wild animal, much as we do with our household pets.

When the animals are as potentially dangerous as the big cats, however, such intimacy cuts two ways. One of the first requirements for a successful trainer of lions or tigers is that he make himself the dominant member of the group. This means that the cats react to him more as the Boss Cat than as a potential predator or—worse—a likely prey. But, since his dominance is never absolute, he must be ever alert for a challenge to his status. He learns to read the behavior of his animals, to anticipate their every move. He gives utmost attention to his position in the ring, especially to the critical distances for eliciting flight or attack. These factors are not only vital to his personal safety but are also an important means of controlling the animal's behavior—of drawing the cat toward him or driving it away, techniques basic to putting the cat through its act.

In spite of his romantic calling, the animal trainer is a practical man. He wants results and knows how to get them. He may not be able to describe all his techniques nor have more than a casual interest in why they work. But the scientific investigation of animal

Learning to live with his charges, a circus trainer stakes his safety on knowledge of their life-styles. Rather than an Indian elephant he chooses an African since for it the tiger, an Asian predator, is not a natural enemy. Patience and companionship have won over the cat, unused to social living in the wild. With courage and psychology Gunther Gebel-Williams builds a pyramid of wonder.

ANTHONY BARBOZA FROM "ESQUIRE"

intelligence and animal learning differs from this pragmatic approach. Such studies spell out some of the conditions involved in effective learning.

Animals are specialists. Each species has its own ecological niche, its own particular corner of the world and way of "doing business." Such specialization is an obvious consequence of the fundamental process of biological adaptation. The kinds of things animals learn, the ways they learn, and the limits of their learning abilities are profoundly influenced by conditions under which they have evolved and in which they live.

The message is clear. To understand how animals learn to live, a knowledge of their natural history is indispensable. This is one of the reasons that field studies are a necessary complement to laboratory research. We want to know not only how an animal learns but why. What is learned? When? Under which conditions?

A logical place to start looking for answers is with the newborn. Here we can see most plainly the intricate and subtle interplay between the forces of heredity and experience. Here we can appreciate most fully why our efforts to force all behavior into neat categories, such as "instinctive and learned," "innate and acquired," were foredoomed to fail. Here we can arrive at a clearer view of what is involved in learning how to live.

No animal, including man, is completely helpless at birth, but the young of most birds and mammals and even some fish can survive only with the active cooperation of the parent. As we would expect, all dependent species have evolved special mechanisms to guarantee that an effective working relationship happens early in life—and quickly. The behavioral patterns often are present in the newborn, but even the simpler ones are influenced by learning. Hand-reared birds learn to gape at the forceps that are used to feed them; each piglet in a litter recognizes and seeks out its preferred teat.

Dependent young frequently develop a strong emotional tie to the parent. This implies that the animal has learned to recognize specific parental characteristics—a learning *of* the parent, we could say. The bond also creates a wealth of opportunities for learning *from* the parent. Finally, the attachment can have long-range influences on relations with others. Most of these factors have been considered in the study of imprinting, without doubt the most familiar means of establishing a powerful filial bond (page 45).

Much has been learned about imprinting in recent years. Konrad Lorenz's suggestion of the 1930's that this close bond could occur only during a very restricted time early in the animal's life, the "critical period," has been generally confirmed—although it is clear that the time is less sharply limited than was first supposed. His idea that imprinting can have an important influence on the animal's recognition of its own species—and later choice of companions and mate—has also been supported. Here too, though, some qualifications are in order. Mate selection very much depends upon experiences after imprinting—whether the animal continues to be exposed only to objects of the type on which it was imprinted, is raised in isolation, or grows up with members of its own species. The outcome can sometimes be bizarre. Take the case of an American bittern that imprinted on people but was able nevertheless to establish a pair relationship with a female bittern. Except that whenever his preferred human appeared, the bird would drive his mate from the nest and solicit the person to enter.

The striking fact about imprinting is that the object of attachment is learned. Some

In the golden aura of joy that envelops master and pet, man reaps rewards from the adaptability displayed by that expert learner, the dog. Affection, loyalty, cooperation, the desire to please—such traits stem from the dog's talent for group living. Inbreeding has heightened the qualities of this long-domesticated animal, but much of its behavior reveals a heritage from the ancestral wolf.

objects prompt imprinting better than others. Early observations suggested that chicks and ducklings were nonselective since they would imprint on objects as small as a matchbox or as large as a person. Later experiments found that items which move, flash, or in some other way provoke attention tend to be more effective than inconspicuous ones. There are also strong inborn preferences for specific patterns of stimulation.

The infant's ties to its parents usually develop more slowly in mammals than in precocial birds, which are up and doing within hours after birth. Lambs can become attached to people, guinea pigs to tennis balls, or monkeys to bath towels if exposure is early, exclusive, and prolonged. Yet there is reason to believe that mammals have unlearned preferences. For example, a baby rhesus monkey vastly prefers as a "mother" a cloth-covered cylinder to a wire one equipped with a bottle from which the infant can nurse (page 47).

The tie to the mother is only part of the picture, of course; typically there is a reciprocal commitment from her. In some species the mother learns to recognize her offspring with the same rapidity that the young is imprinted on the parent. The mother goat needs only a few minutes of nuzzling and licking her newly born kid to identify it. If at birth the kid is

DURWARD L. ALLEN. OPPOSITE: L. DAVID MECH

Tracks of a wolf pack stitch the white coverlet of Isle Royale in Lake Superior. This healthy moose will escape; usually only the weak fall to the fangs. Hunting tactics of the island's gray wolf (above) and of wolves elsewhere may reflect social tradition: Sensing prey, they often rush to touch noses; the ritual, akin to pups' food-begging, may tell the leader the pack is hungry and thus stir him to lead an attack.

separated from her and then returned in an hour or so, she will usually treat it as a stranger, butting it away and refusing to let it suckle.

We can see then that parent and young are a functioning social unit. The behavior of one cannot be appreciated without noting the behavior of the other. In the presence of the parent, the infant shows comfort and contentment. If contact is lost, the result is instant distress. Nearly always, distress is conveyed in unmistakable terms: crying, chirping, whining, mewing, bleating—sounds that evoke tender impulses in man, as we must assume they do in the species for which they are intended.

I recall an incident that occurred while I was studying *Callicebus* monkeys in South America. Among these primates, attentive mothers often allow their babies to be carried by other monkeys. On this occasion I had taken into the forest a baby *Callicebus* that I was hand-rearing. It was so small I carried it concealed in a shoulder bag, where it was usually quite content. But now, perhaps hungry, it gave the distinctive call of a distressed baby *Callicebus*. Thereupon a female in the trees above me, which had left her infant on another monkey's back, rushed to her offspring. She retrieved the baby, inspected it, then clasped it to her breast. Apparently she had mistaken the cries of my infant for those of her own.

The intense, two-way social bond between mother and young fulfills some of the basic conditions for effective learning. The mother becomes a strong motivator of infant behavior; one has only to see how a young monkey or gosling scrambles to keep up with its parent to be convinced of this. She also is a potent generator of reward and punishment. And finally she is a rich source of sensory information or cues. What the immature animal learns from its parent and how such learning is brought about depend very much on the species and the circumstances. Chicks, for obvious reasons, do not learn the same things as wolf cubs or monkeys. And it is just as obvious that the monkey brought up in a crowded Delhi bazaar has an entirely different educational prospect than one bred in the forest.

379

One of the most primitive and widespread forms of social influence is called social facilitation, which essentially can be said to occur whenever a specific activity by one individual brings on or steps up a similar activity in another. A favorite example in man is the chain reaction that follows a yawn. Now, I have never seen a monkey or an ape by its example induce another to yawn. Nor have I been able to get one to yawn myself. But this doesn't mean that man cannot cause social facilitation in animals, although it does show that susceptibility is selective. There is no question that for chimpanzees a human example can be compelling. Robert M. Yerkes, for instance, was able to get chimps to accept, chew, and sometimes even swallow tasteless filter paper by pretending—noisily and with evident satisfaction—to eat it himself.

Social facilitation takes many forms. Ants work harder in the presence of other ants. Rats become sexually aroused when allowed to see copulating rats. *Callicebus* monkeys can be induced to call by the sound of another monkey calling. There are times, moreover, when an animal not only is stimulated by the example of another to do something it already knows how to do but will also actually learn something new just by watching. This imitation, or observational learning, is another process through which parents can educate their young. Obviously there is an enormous biological advantage in acquiring a skill or learning the consequences of an act at a distance since it reduces the effort, as well as the risks, entailed in trial-and-error learning.

Ability to learn by observation is widespread among warm-blooded animals, but it is by no means as generalized a trait as one might suppose. Even among animals that have the best reputation for imitation, some things are learned by observation more readily than others. "Monkey see, monkey do" is true only for some monkeys and some actions.

When we look at observational learning as it occurs in nature, we usually find a complex intermingling of the old and the new. Hardly any animal starts from scratch. Jane Goodall's young chimps watch while adult chimpanzees break off twigs, strip away the leaves, and fish in termite holes. The entire sequence is long and fairly elaborate and it is doubtful that many individuals would become skilled termite fishers if they never had an

Deft arms pop a cork for a bottled lunch. To get rewards of shrimp, the octopus—perhaps the most intelligent of all sea dwellers save marine mammals—first learned to enter the jar through the uncorked hole. Then it mastered knocking aside a stopper loosely put in place. Ultimately it could drape itself over a jammed-in plug, rock it out with sucker-equipped arms, and slip inside. (Octopuses a foot long with bodies as big as tennis balls can squeeze through half-inch holes.) Studies show these brainy creatures can discriminate shapes, thread mazes, and remember events.

380

opportunity to observe other animals at work. At the same time, however, there are some parts of termite fishing behavior which are commonplace; all young chimps are fond of playing with sticks and like to poke them into holes. We can assume that, having cadged a few termites from an older animal and noted how they are taken, the young chimp is able to put his rudimentary skills into a smooth and effective performance. And perhaps this is a more useful strategy than a highly generalized tendency to imitate—which, carried to an extreme, could result in a witless and compulsive parroting of someone else's tune.

In a sense the older animal is serving as a teacher. Usually, however, a teacher is more than a model; he adjusts his behavior to the needs of the pupil and persists in this effort until the student achieves a certain standard of performance. This is a very difficult proposition to demonstrate in creatures other than man.

Perhaps the closest we can come to acceptable examples of teaching occur in the cat family and among certain primates. For instance, George Schaller reports that a mother cheetah once brought a live gazelle fawn to her 5-month-old cubs and released it. Only after the cubs had made several ineffectual attacks did she intervene and make the kill. And when one recalls the difficulties that the Adamsons encountered in training their home-reared lioness Elsa to hunt, it is easy to believe that the mother cheetah was providing an important learning opportunity for her cubs.

What we can state with complete confidence is that many animal parents interact with their young in ways which bring about abiding changes in behavior. Take weaning. It usually requires some force—and occasions some protest—before it is accomplished. This is only the most common of many examples in which punishment or some form of physical deterrent is used to eliminate unacceptable actions. In certain cases the element of self-interest is less obvious. A mother macaque monkey will prevent her infant from eating an unfamiliar food. Older baboons—not necessarily mothers, for it is also done by males—have been seen to station themselves in front of a baited trap and drive off or grab any young member of the band that tries to enter.

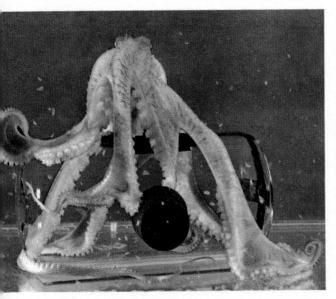

Reward for an act has just the opposite effect from punishment. Instead of suppressing the act that it follows, reward tends to strengthen or reinforce the act. Reacting promptly to her infant in distress, the mother increases the likelihood that it will cry out on similar occasions in the future. Possibly this explains why chimps raised without a mother so seldom scream or whimper when they are upset.

Now let us broaden our horizons somewhat. Where an animal lives—and what animals it lives with—play an essential part in its education. In the more gregarious, group-living species—the ungulates, many rodents, and in particular the primates—the mother only begins the learning process. Every individual that the growing animal comes into contact with can contribute something to its education. Young prairie dogs learn the boundaries of their home territory because they are attacked whenever they enter the territory of a neighboring group. Young rats discover that they are treated roughly when they approach adults too closely at a food source. And the endless rough-and-tumble play that is characteristic of so many young animals probably helps sharpen fighting skills.

Gradually the growing individual enters into stable relations with other members of its group. Friendships and coalitions develop. Differences in social rank are established. And there emerges the basic *we/they* division—the ancient, invidious, and seemingly inevitable distinction between the ins and the outs—that is found in so many group-living animals and that, for man, has been the source of so much unhappiness and strife. How this distinction is made varies in different species. Among insects and many rodents, recognition of a fellow group member depends upon a distinctive odor; in birds and nonhuman primates, the critical cues are usually visual. In man, we suspect that distinctions are more often matters of attitudes and values.

Early social experience can profoundly influence even elemental skills involved in mating and care of young. Male sexual performance is impaired in guinea pigs, cats, dogs, and monkeys that have been raised in isolation from their kind. Female chimpanzees and rhesus monkeys that have been deprived of their mothers at birth—and have had no opportunity to deal with babies—make poor mothers themselves, particularly with their firstborn. In nature the female has many opportunities to play mother while growing up with younger animals. Women's lib notwithstanding, adolescent female primates are fascinated by babies. Enduring relationships can develop in which a young female serves as a second mother. It isn't even essential that the "baby" be of the same species. I once had a young female rhesus that adopted a full-grown albino rat. She slept with it cradled in her arms, groomed it, and carried it with her wherever she went. What the rat made of all this I cannot say, but I know he was sleek and fat.

Male chauvinism notwithstanding, it must also be said that infants hold a special attraction for many primate males. In some bands of Japanese monkeys, for example, certain males "adopt" the yearling young, doing nearly everything that a mother would. We find here, then, a clear suggestion of a social tradition, for the practice is a local custom,

Out of the intimacy of parent and young come ingredients for rapid and efficient learning in the animal world. A raccoon mother knits a social bond through the nurturing of her kits; another leads her offspring on a stream-bed search for food. Thus does the experience of one animal become a force that molds another's ways—and learning become a means for adapting a society to change.

shared by many members of the society and handed down from one generation to the next. Tradition, in fact, probably plays a much larger part in the natural life of animals and their societies than is generally supposed.

Places selected for courtship, mating, and the rearing of young are particularly likely to be favored by tradition. The communal display grounds of the prairie chicken are tamped hard by untold generations. As we have seen elsewhere in this book, birds of many species return to the same rookeries year after year to breed and to rear their young; northern fur seals, with a strong traditional attachment to place, return during the breeding season not only to the island on which they were born but often to the same site.

Traditions are transmitted in various ways. In some animals the older generation need do little more than select the place where the next one will be born. We see this in the Pacific salmon which, in one of the marvels of migration, return from distant ocean cruising to the very freshwater creek where the parents spawned and then died (page 279). Clearly the recognition of the home stream is learned, for when salmon eggs are taken from a traditional spawning site and allowed to develop in different waters, it is to the stream where they lived as fry that the adults return.

How do traditions begin? For the most part we do not know. But Japanese scientists have succeeded in giving us a good picture of how traditions can be created and maintained in monkey societies. The investigators concentrated particularly on the social transmission of food habits in their native macaque. Early in their studies they discovered that some feeding habits are traditional. One group of monkeys will dig for roots while another will not; some groups invade rice fields but others never do, even though they have lived many years in a region where paddies abound.

To determine how traditions originate and spread, the scientists offered the monkeys foods not part of their regular diet—caramels, for example. Younger offspring and adolescents were usually the first to adopt the new food, and they passed the habit up the age ladder to their elders. Not all the adults learn to accept the food, but this will not prevent it from becoming an item on the traditional menu, for when the young mature, they transmit the habit down the age ladder to their own offspring. Sometimes there is definite invention in the beginning of a tradition. Two examples: the use of poles as scaling aids by young chimpanzees (page 368) and the potato washing and wheat sifting started by Imo, the Japanese monkey (page 42). Both her innovations were made when she was still young.

That youth can be a major instrument for social change will hardly come as news in today's world. Yet I am intrigued to find so clear a parallel with human experience among creatures whose connection with our contemporary scene is, at best, remote. Among the young, discovery and innovation are most prominent because more options are open to them, they are a less finished product than the more experienced adult, and they are more active, more curious, more willing to venture and explore. While studying monkeys in the field, Emil Menzel placed small toys and other objects beside familiar monkey pathways. He found that passing animals which handled the toys were almost always immature, though he had no doubt from the way the adults behaved that they were fully aware of the novel items. At times, it even seemed that the mature animals were actually waiting

and watching while the juveniles tested the situation. (Continued on page 390)

THE USE OF TOOLS...

A solitary wasp grips a pebble and pounds dirt into a nest burrow. An elephant scratches its back with a stick wielded by its trunk. Gorillas and chimps throw branches at intruders. In each case an animal has learned to use a tool to extend its body's capabilities.

In the Galapagos Islands the woodpecker finch, most remarkable of Darwin's famed finches (page 185), burrows in trees for meaty grubs. But lacking a woodpecker's long barbed tongue, this amazing artisan pries out its meal with a cactus spine. A researcher saw the skill develop in a captive young finch during manipulative play with twigs. At first the bird dropped a twig and probed for insects with its beak, often in vain; in time the bird held on to the twig and began to pry with it. Another finch, stymied by a forked twig that wouldn't work, snapped the fork to fashion a proper tool.

385

...to crack an egg

Heads high, stones in their beaks, Egyptian vultures zero in on an ostrich egg, its shell a sixteenth of an inch thick—too tough to peck open. Downward snaps of the neck propel the missiles. About half the time the stones miss the target (below left). After a few hits the shell breaks; as the yolk begins to spill, another Egyptian vulture and a darker hooded vulture rush over to join the feast.

Observing the tool-using vultures during a National Geographic Society expedition in Tanzania, Jane Goodall saw them range as far as 50 yards from a target for ammunition that weighed up to 18 ounces. They bombarded any egg they couldn't pick up, even red or green decoys. One large fake they pelted for 90 minutes. But a white cube the size of an ostrich egg they ignored. Thus shape may be the major factor that stimulates stone throwing.

The use of stone tools seems to derive from the Egyptian vulture's habit of flinging small eggs to the ground to crack them—the bird's movements in both activities are identical. But stone throwing is not strictly inborn. One young vulture pecked at an ostrich egg for 30 minutes without success. Then an even younger bird moved in, picked up a stone, and fractured the egg in 6 minutes—suggesting that each bird must acquire the skill for itself.

HUGO VAN LAWICK

...and shuck a shellfish

Face averted, a California sea otter smashes a shellfish against a rock "anvil" balanced atop its stomach. Sea otters also use stones to knock loose abalones like the big one being hauled up. The heavy work done, a successful hunter samples abalone on the half shell.

The sea otter employs tools more frequently than any other mammal except man. One energetic otter brought up 54 mussels in 86 minutes, whacking them against a stone a total of 2,237 times! The species shows evidence of an inborn tendency to pound objects; youngsters may learn actual tool use by watching their mothers feed.

Such notable skills are still far from those of early man. "Though other animals share with man tool-using and to a minor extent tool-making ability," wrote George Schaller, "there still appears to be a wide mental gap between preparing a simple twig for immediate use and shaping a stone for a particular purpose a day or two hence."

JAMES A MATTISON, JR. UPPER LEFT: BATES LITTLEHALES, NATIONAL GEOGRAPHIC PHOTOGRAPHER

It is easy to see how both the liberal and the conservative responses to change serve useful and necessary functions within animal societies. The biological wisdom of conservatism is obvious. Why tempt fate? Why chance unleashing a demon until the need clearly justifies the risk, until the benefits are no longer tainted by the slightest doubt?

The problem, of course, is that by the time certainty is achieved, the twelfth hour has come and gone. And so the biological wisdom of innovation is just as plain. The world has ever been a changing scene. One can attempt to make it otherwise only by retreating into those small corners where the conditions of life seem forever fixed. But they never are, really. Fossils record the history of creatures that have failed their own great challenge to change and so passed into oblivion. Adapting to change when necessary is what evolution is all about. And the ability to learn is first and foremost a way of dealing with change. As a teacher, I am encouraged to believe that the young have a special part in the process of discovery and innovation that learning makes possible. Theirs is a high-risk occupation, to be sure, yet one that no viable society can afford to be without.

But it is fully as important to remind ourselves that the ability to learn is not tied to any single age group or social class. Learning to live is a continuous and never-ending process, and some degree of flexibility for it is always preserved. The ability to learn, itself the product of organic evolution, thus opens up a whole new way of adapting to change, one far more rapid and flexible and efficient than the genetic modifications that are the recognized vehicles of evolutionary advance.

As we have seen, the fact that man must learn how to live in no way distinguishes him from many other animals. What distinguishes man is the remarkable degree to which he has come to rely upon learning. Our concern with the process is plain to see. We analyze learning in the laboratory; we design our children's toys so that they will not only amuse but instruct; and we have created cadres of experts in the learning arts. *Homo sapiens,* we call ourselves. Though at times it seems we have been a little hasty, a shade arrogant perhaps, to claim wisdom for ourselves alone, there is no denying that our credentials are impressive. Most other animals are specialized learners. Some of their accomplishments are remarkable, of course. The rat has demonstrated its superiority to the college sophomore in mastering the intricacies of the tunnel maze, and the skill of the average pigeon in winging unerringly home would put many an Explorer Scout to shame. But these are specialized achievements by specialized learners, closely tied to their problems of everyday life. Man is not a specialized learner; he is a learning specialist.

So powerful a force has human learning ability become that it enables us to alter the face of the earth in the space of a few generations. In our brief history we have solved many of the problems that the natural environment imposes, and in so doing we have created new ones in their place. Now we are beginning to wonder how we can make the best use of this formidable talent. Who shall be master, and who shall serve? Can we learn to live with our marvelous ability to learn?

With a waving flag their Pied Piper, ducks waddle a dawnlit lane on the island of Bali; all day they will forage within sight of the pole stuck in a rice field. Though trained to the cloth, the fowl are born with a follow-mother survival trait so strong it can induce them to pursue the first
object they see. This process of imprinting is one of the marvels of learning that shape animal lives.

GILBERT M. GROSVENOR, NATIONAL GEOGRAPHIC STAFF

EPILOGUE:

QUEST FOR THE ROOTS OF SOCIETY

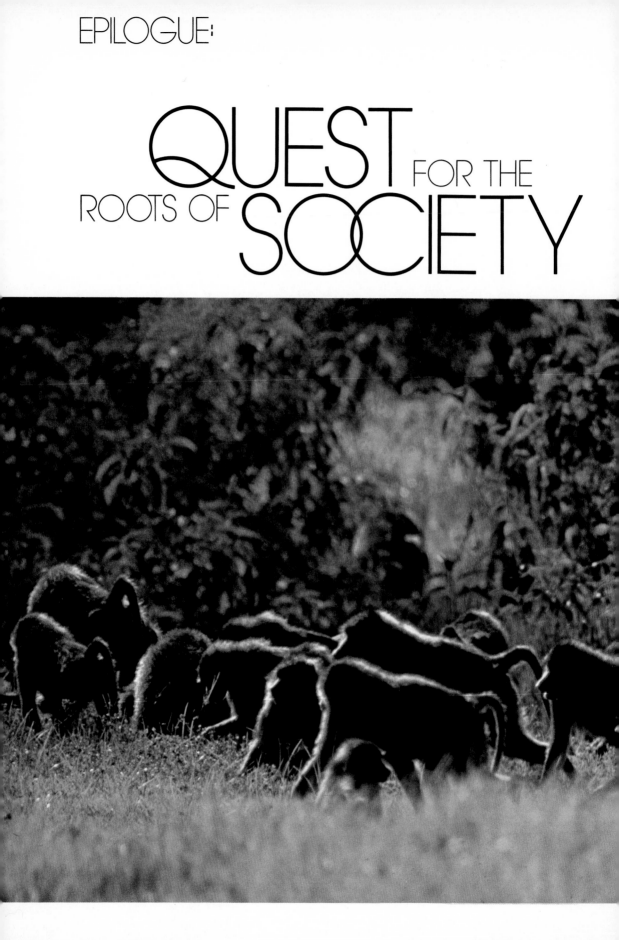

Nairobi National Park, July 5 (from field notes): 4:55 P.M. — My wife Nancy and I drive east toward the edge of the park to check on a baboon troop we spotted through binoculars. We keep the Land-Rover in low gear, creeping nearer. It's the Python Pool troop—there's the male with only a stub of a tail. The group of 47 feed leisurely, moving toward their sleeping trees in Hyrax Valley, a shallow ravine. This is one of our favorite sections of the park, separated by a gentle rise from the main tourist road. On our left a herd of wildebeests snort and stamp. Ahead, six ostriches move away slowly, pivoting their heads to keep us in view. . . . For nearly an hour the baboons feed. They reach a water hole and linger, not moving on to the trees. Why, I wonder?

5:50 — It is clear the baboons are quite disturbed. The troop is unnaturally compact, clustered near the adult males. My first thought is that they are nervous because lions have

BABOONS, ENTERING UNDERGROWTH WHERE PREDATORS MAY LURK, GROUP FOR DEFENSE. M. PHILIP KAHL

recently been here. Some Grant's gazelles trot skittishly back and forth, and birds, invisible in the thick foliage, give alarm calls. A juvenile baboon goes up to a grizzled, nearly toothless male, snuggles up briefly, and grooms the old male's chest.

5:55 — Three adult males and two females stare down into Hyrax Valley. Several infants give the *ick-ick* and *ooing* sounds of unrest. Juveniles moan softly, huddled against adults. Something in the valley is keeping the troop from the safety of the sleeping trees. We scan the valley with binoculars but cannot see the cause. Night on the Equator always comes swiftly, and as the time nears, the baboons' distress grows.

6:00 — Several adult males pace rapidly back and forth. Suddenly, the grizzled old male gives three deep-throated, authoritative grunts and strides purposefully down the faint game trail toward the trees. Other adult males join him. Younger males overtake and dash past him at full gallop. Sighting over their backs, we at last make out the top of a cheetah's head, peering over tall grass. Now the rest of the troop advances, infants hanging onto their mothers' backs, keeping well behind the charging males. All seven adult males rush the cheetah at top speed. Suddenly the cheetah stands — and there's a second cheetah. . . . Now the baboon males are about 20 yards off. . . . Both cheetahs dash away in graceful leaps, clearing a stream and coursing up the far bank. As they near the crest of a hill 100 yards away, we see a third cheetah has joined them.

So intent were we on the chase we did not notice when the seven older juvenile and adolescent males ran forward and climbed a low tree. But there they all sit with a balcony view of the action, their eyes riveted on the attacking adult males.

6:07 — The chase not yet over, females resume feeding with what seems to us great nonchalance. Then infants again hop aboard, and the troop moves toward the trees.

6:15 — The baboons are near the trees, but several males still walk purposefully up the far side of the valley. The cheetahs are still in retreat about 100 yards ahead. The males' *bra-hoos* resound after them so long as the predators are in view.

6:24 — Some of the baboons begin to climb into the 20-foot-high trees in the valley.

6:30 — It's almost completely dark. The Python Pool troop is settled in for the night.

In our seasons of watching baboons, including a year in 1959-60 and the summers of 1963 and 1964, I've seen them near large predators less than a dozen times. Yet encounters with predators are the key to the social organization of African savanna baboons. Though clashes are rare, each is a crucial test and it becomes part of a troop's tradition.

The baboons of Kenya are part of a widely distributed and numerically successful group of primates. There are five kinds of baboons. One, the gelada, is distinct and today is found only in the highlands of Ethiopia (page 306). The other four are closely related. The hamadryas, or sacred, baboon of Egypt is adapted to semi-desert conditions. Two others, the harlequin-hued mandrill and its drab cousin the drill, are adapted to the jungle and live in coastal regions of West Africa. The fourth, the common savanna baboon, is distributed throughout the savanna regions of Africa, from Dakar on the west coast to Mombasa on the east, and south to Cape Town.

Like man or any other animal distributed over such vast distances, the savanna baboon exhibits local variations. In West Africa, for example, it is small and reddish; in Central and East Africa it is larger, more robust, and olive green in color. All varieties are considered members of one species. From a series of studies we know that in basic behavior and social organization all savanna baboons are comparable.

African baboons and the closely related macaques of Asia represent a widespread evolutionary radiation of Old World monkeys that, like man, have adapted to life on the ground. Their behavior offers useful clues to the earliest stages of human evolution — that is, hominid evolution at the time our ancestors became distinct from the ape line.

The basic home of all primates — prosimians, monkeys, and apes — is the forest. Primates have been adapted to an arboreal life for more than 40 million years, and among about

Order in baboon society rests on an understanding of who's who in the hierarchy. Top males defend the troop from predators and also police domestic affairs, as when rowdy juveniles disturb their elders. Young males vie for position and sharpen battle skills by fighting each other. Two charge into a fray (far left). Another one flees (center). A dominant male onlooker slaps the ground, signaling "Enough!" When his command fails to break up the fracas, he intervenes bodily to restore peace in the ranks.

IRVEN DEVORE

600 living varieties only a few can wander far from the safety of trees. For millions of years the ground was a niche open to primate exploitation. In the New World, none of the many monkey species has ever made the transition that the baboon-macaques made so successfully in the Old. Man's apelike ancestor was able to make a parallel transition.

In terms of genetic kinship, man stands much closer to the great apes—especially the chimpanzee and gorilla—than to the baboon. But in terms of ecology, our hominid ancestors, facing similar problems, may well have shared some important adaptations with these monkeys which left the trees millions of years ago and moved into open country.

Major differences in life-style would confront any primate attempting such a change. The forest is, after all, a safe refuge. Scampering through the interlocking branches of the canopy, monkeys and apes can elude most predators. Efficient, grasping hands and feet; sharp stereoscopic, color-discriminating vision; and capacious brains equip them to escape almost any danger. By using all four feet (and in the New World, a tail) to grasp and divide their weight on small branches, monkeys and apes can usually elude big tree-climbing cats and snakes. If an eagle or a hawk threatens from above, they can drop to lower branches. In fact, the arboreal way of life has worked so well that various monkeys and apes of New World, African, and Asian jungles have achieved combined population densities of several hundred per square mile, matching those of man in many urban areas.

Safe as it is, jungle life can be dull. Forest monkeys interact with few other animals: primarily birds, squirrels, and other monkey species. By contrast, baboons in a day may meet 30 species of mammals alone—ranging in size from mouse to elephant. And on the savanna they are the potential prey of a wider variety of animals, especially the large cats—cheetahs, leopards, and lions. A young, crippled, or sick baboon is also vulnerable to eagles, hawks, pythons, jackals, and hyenas. (Continued on page 401)

AN ADULT FEMALE BABOON GROOMS A DOMINANT MALE. IRVEN DEVORE

PRIMATES' PROGRESS

Success in evolution means numbers—and the macaques are eminently successful animals. With the closely related baboons, these ground-dwelling monkeys have adapted to more of the earth's surface than any other primate but man. In fact, their combined range, from Gibraltar to Japan, almost exactly matches that of early man.

In India, populations of one macaque species, the common rhesus (above), have been estimated in the millions. Adapted to city, village, temple, forest, and field—and popularly regarded as sacred—they had the run of the realm for centuries. The langur, a tree-dweller, was actually the sacred monkey of Hindu legend, but pious villagers pampered the rhesus as well. No longer tolerant of monkey mischief, present-day Indians shoo the too-prevalent primates from gardens and bazaars. Each month thousands of rhesus juveniles are trapped for export to laboratories. The Rh (rhesus) factor in human blood is one lifesaving discovery in which they played a part.

Simians in snow country

Snowbound Shiga Heights northwest of Tokyo seems too harsh a world for monkeys—animals that originated in the tropics. But as earth's climate turned colder ages ago, Japan's hardy macaques adapted and survived. Today they range the farthest north of any nonhuman primate, to Shimokita Peninsula at about the latitude of Cape Cod. Enduring frigid weather four or five months a year, they ball up in thick coats to keep warm while resting. Stumpy tails may be in part an adaptation to cold; long tails freeze and drop off.

Snow monkeys eke out a winter living by eating bark and the needles of evergreens and by digging grass from beneath the drifts. They bear young later in spring than macaques of the same species in southern Japan; even so, mortality runs high and troops are smaller. For all its bleakness, Shiga Heights offers one luxury—a hot spring, where venturesome monkeys can enjoy a daily splash.

Life with mother

"Apes esteem their young the handsomest in the world." Rabelais penned it—Flo's family proves it. The infant Flint is mother Flo's chief interest and his sister Fifi's "doll." She lavishes attention on her baby brother, even kidnaping him for short periods.

Chronicling free-living chimps in Tanzania, Jane Goodall found mother-offspring groups the most stable social unit. Chimps stay near trees; males, not needed for defense or tied to infant care, range farthest, at times killing game. Rudimentary role division, tool use, and grudging sharing of meat by these advanced primates foreshadow traits held basic to human social evolution.

Danger has a profound effect on the organization of primate society. So long as monkeys or apes are living in the trees, the best defense is flight. There the rule seems to be "Every monkey for himself and predator take the hindmost." In open country, flight often is not feasible; among primates it is no accident that only in baboons do we find all adult males continuously associating with the females and offspring in a cohesive social group. On the savanna the adult male baboons are the troop's only protection, and they are superbly equipped. The male weighs up to 100 pounds, twice as much as the female. He possesses enormous, daggerlike canine teeth which he constantly hones against specialized pre-molars. An average troop of 40 baboons has about 7 adult males; any 3 of them can put to rout all predators except a pride of lions.

Our hominid ancestors solved the defense problem by developing tools and weapons. True, the human species displays a certain amount of sexual dimorphism. The difference in height and physical robustness between men and women is of about the same order as in chimpanzees. Human males especially differ from females in the strength of the "fighting muscles," such as biceps. Though the human solution to the problem of group defense is mostly cultural, in both baboons and man it is the males who wield the weapons.

For baboons, defense means to incorporate in the group animals that are aggressive by nature and physically capable of killing a cheetah or leopard. The social solution is a familiar one: the dominance hierarchy. Each baboon in the troop knows his or her position. Aggressive acts travel downward through the ranks. For example, an adult male chases a less dominant male; he chases an adult female, and she chases a juvenile.

Communication often substitutes for fighting. A baboon's repertoire of sounds and gestures can convey various degrees of threat or appeasement. To warn another baboon of his intention to attack, an adult male may first stare at the offender, next slap the ground where he sits, or he may stand, erecting his ruff. He may show the white above his eyelids as he opens his jaws in full display of his canines. If the opponent does not flee or show submission, the aggressor charges, roaring and grunting loudly.

A MALE BABOON'S DISPLAY OF CANINES MAY SIGNAL THREAT, FRUSTRATION—OR MERELY FATIGUE. IRVEN DEVORE

One of our most interesting findings was that, of adult males which head the dominance hierarchy, two or three usually stand together in a core group of "central males." These males are often past their physical prime, yet they dominate younger, stronger males because they support each other when fighting breaks out. By forming a coalition—a baboon "Establishment"—older males can both suppress disruptive striving for dominance by young males in the group and remain vigilant toward danger from the outside.

Like his human counterpart, the adolescent baboon male experiences years of frustration between physical maturity and social maturity. As an adolescent, he tries to develop dominance over adult females in the troop. Adult males continually thwart his efforts by aiding the female, driving the young male from her. Constantly at odds with the Establishment, he is temperamentally bold; he often makes hundred-yard detours from the group to sample fruit from a tree, or from the safety of a high branch teases a predator. By human standards, it would be easy to call him delinquent. Paradoxically, this bold behavior makes for male success and may benefit the troop. Constant fighting hones skills of attack and defense. In a crisis, a male that knows the terrain may lead the group to a safer place.

Central males, meanwhile, tend to subdue the younger male until he is seasoned by age and experience and more likely to act in ways that will protect females and young.

The life experience of baboon females is very different. At only a few weeks of age, males play longer and rougher; and by age 3 when males begin to fight more seriously, females stay in quieter zones around adult females and infants. Although all baboons are tolerant of infants at play, young females take every opportunity to touch, fondle, and hold them. There is every reason to believe that effective motherhood requires practice. Primates are said to have a "social brain"—social skills can only be learned in a social context. And while many mammals have frequent, large litters, a female primate has only a few offspring in her life. Early practice in infant care helps ensure their survival.

Sexual behavior in all nonhuman primates is limited to the female's estrous period, 5 to 15 days each month. During the early days of estrus the female baboon is mounted by juveniles, adolescents, and young adult males. As her receptivity increases, she mates with higher-ranking males. At the peak of estrus—when she is most likely to conceive— she mates mainly with central males. Thus it is not young adult males at their most vicious,

Dropouts from tree life and not so agile as arboreal monkeys, savanna baboons can still leap nimbly when they must. Members of a troop in Serengeti National Park take a limb-to-limb route over a stream (left). Like many African animals, baboons do not like to go into water; those that did may have been selected out by hungry crocodiles. A fight in the trees, where baboons sleep to avoid nocturnal predators, may cause the loser to risk a daring leap.

Contentious males rush at each other, slashing viciously with their canines. Bested, the older male hunched at right screams. Cries often bring help; serious injury is seldom inflicted.

403

RICHARD D. ESTES, PHOTO RESEARCHERS. ABOVE: IRVEN DEVORE

Tense and combat-ready, baboons in close array hurry through a danger zone. Highest-ranking males escort females and infants at center. Other males in their prime guard front and rear.

Massed males meet a formidable foe

Sharp, two-phase warning barks from one of the males in the van trigger troop defenses. Freezing momentarily, members drop back a few feet awaiting cues from the leaders: Is it flee or fight? Battlewise and responsible central males rush forward to assess the danger. From tall grass springs a leopard. Now numbers count; even leopards may be losers. While vulnerable troop members—females, infants, and young juveniles—flee, a fighting phalanx of adult males closes in on the adversary. Such confrontations with predators, though rare, reveal the reason for the baboons' social unit. Coming down from trees or cliffs where they sleep, they embark on a daily food trek of three miles or more. While foraging in open country, the troop depends for survival on the muscular males with their slashing canines.

"There just isn't a life for a baboon apart from the group," the author comments. "If this leopard isn't cornered, I would predict that in the next instant it will turn and run like a scalded cat. Its strategy is to try and grab a female, a cripple, or a juvenile and be off with it before the rest know what's happened. No leopard in its right mind wants to face these adult and subadult males—it's just not worth it."

DRAWINGS BY GEORGE FOUNDS

aggressive stage that reproduce most often but older males at the stage when they are
heading coalitions, protecting weaker animals in the group, and chasing away predators.

The female monkey or ape does not come into estrus during pregnancy and lactation;
thus mating does not overlap the maternal role. Sexual activity is confined to mere weeks
of her adult lifetime. No long-term pair bonds form on the basis of sexual attraction.

Scientists formerly assumed that the female's major role in the primate group was bearing and caring for infants. But long-range studies of rhesus macaques and chimpanzees
show that the female may continue to have great influence over her offspring into early
adulthood. Mother-offspring subgroups may remain together for many years. No study of
savanna baboons has continued long enough to be certain that such uterine kin groups are
important in this species. We do know, however, that baboon males, like macaques and
chimps, may change from one group to another while females almost never do.

Among nonhuman primates we do not find an exact counterpart of the human family.
But in baboons and other species we do find kin groups related through the mother, adult
males attentive to the young, and a complex troop organization—patterns of social behavior that are part of human family and tribal structure.

M an, the human primate—what features began to set his society apart? Among the
most important are tool use; the home base, or fixed camp; the family, with its
distinct sexual pattern; an economic division of labor based on sharing; and language.
Recent archeological evidence reveals that our ancestors used crude "pebble tools" at
least three million years ago. We have only inconclusive evidence for the earliest appearance of the home base, from which an economic division of labor could have arisen.
Excavations still in progress should soon clarify this crucial development in hominid
society. The advantage of a home base is obvious from studies of living hunter-gatherers:
The men can go hunting while the women and children gather vegetable foods nearer by,
knowing all will return to camp to share the results of their labors.

We have begun to see ways in which this fundamental economic change in hominid
society could have evolved. We now know that chimpanzees use simple tools: crumpled-
leaf sponges to sop water and twigs to probe for termites. We know that a female baboon
or chimp could easily gather enough food on most days for herself, her offspring, and an

Meandering along an open road
in Nairobi National Park, a troop
of olive baboons takes its ease
(opposite). Some members groom,
a ceaseless pastime that cuts
across hierarchical lines.
Some dig roots with their hands;
others overturn rocks seeking
ant eggs and larvae, a delicacy.
Voracious grass-eaters, baboons
can digest some 200 varieties
of plants and roots, which they
rub free of grit to save molars.
On occasion they discover and
kill a small mammal or bird;
ranking males usually eat the meat.

Rounding a bend, the troop
meets the unexpected: A dummy
leopard—animated by a small
motor—has been placed in the
path by experimenters to test
the troop's response to a predator.
After animals in the forefront
voice a warning, older males
come forward to reconnoiter.
It looks like a leopard. It moves
but does not attack. Other males
join the leaders. They dash
up to the decoy, then—as if
baffled—mill at roadside,
working themselves into a frenzy.
At last, the central males
begin to maul the leopard,
tearing at the abdominal area.
BOB CAMPBELL

adult male—but she doesn't. And it is the males of these species that are best at capturing game. But without the reinforcing behavior pattern of sharing, an economic division of labor has not developed. Among nonhuman primates, only chimps may reluctantly give bits of meat to an individual begging for it, a step toward the human system.

Hunting in itself is an insecure economic base. Recent studies of living hunter-gatherers, such as the Bushmen of the Kalahari Desert, reveal that men can go hunting only because they can rely on significant food contributions by the women—the source of 50 to 80 percent of the Bushmen's daily diet. This indicates that hunting could not become a principal feature of human social evolution until males and females shared food.

Once sharing was established, the family could become a viable economic unit, with continuous pair bonding of male and female and cooperative rearing of children. Pair bonding must have increased pressure for females to become continuously receptive. Constant sexuality seems fundamental to a family system, but we can be relatively certain it evolved only after the pivotal economic change: It served the economic system.

One day on the savanna I was away from my truck watching a baboon troop when a young juvenile came and picked up my binoculars. I knew if the glasses disappeared into the troop they'd be lost, so I grabbed them back. The juvenile screamed. Immediately every adult male in the troop rushed at me—I realized what a cornered leopard must feel like. The truck was 30 or 40 feet away. I had to face the males. I started smacking my lips very loudly, a gesture that says as strongly as a baboon can, "I mean you no harm." The males came charging up, growling, snarling, showing their teeth. Right in front of me they halted, cocked their heads to one side—and started lip-smacking back to me.

They lip-smacked. I lip-smacked. "I mean you no harm." "I mean *you* no harm."

It was, in retrospect, a marvelous conversation. But while my lips talked baboon, my feet edged me toward the truck until I could leap inside and close the door.

It is in language that man most clearly differs from his primate relatives. Baboons and chimpanzees use many expressions of threat and appeasement. Chimps also have elaborate greeting behaviors: their versions of shaking hands, embracing, kissing. But only man, so far as we know, has a communication system that encompasses the past, the future, and what lies far beyond his own horizons. All animals lack the phonemic principle found in every human language—the ability we have to combine, and readily recombine, bits of meaningless sound into words and sentences. Recent experiments in which chimpanzees have been taught to use symbols reveal unsuspected subtleties of the chimp brain. But these animals are "linguistic" only because of human tutelage; their skills play no part in the communication system of chimps in a natural environment.

Fossils alone cannot reveal when language became part of the human repertoire. But the hunting hominids must have found language advantageous in planning their forays. Once our ancestors acquired language, they could elaborate all forms of human behavior and institutions. Language, in turn, made possible the entire symbolic world of the human mind: the worlds of art, music, religion, mythology, and history—and such pursuits as the study of animal behavior.

Baboon mother and nestling infant evoke a primal bond that undergirds societies of animal and man.
STANLEY WASHBURN

BIOGRAPHICAL AND REFERENCE NOTES

MAN AND ANIMAL

Leonard Carmichael, who died in 1973, was Chairman of the Committee for Research and Exploration of the National Geographic Society, whose grants helped support the scientific work of authors Schaller, Payne, Eisenberg, Fossey, Estes, Kahl, Lott, and Riopelle.

Dr. Carmichael, who received his Ph.D. from Harvard, served as president of Tufts University from 1938 to 1952 and as Secretary of the Smithsonian Institution from 1953 to 1964. He was president, from 1961 to 1969, of the Section of Experimental Psychology and Animal Behavior of the International Union of Biological Sciences. He was a member of the National Academy of Sciences and president of the American Philosophical Society, the Nation's oldest learned academy. He published widely on prenatal behavior in mammals. Suggested additional reading: *Ethology: The Biology of Behavior* by Irenäus Eibl-Eibesfeldt.

THE DRIVE TO SURVIVE

Peter Marler, professor at Rockefeller University, was born and educated in England. He obtained a Ph.D. in botany at the University of London and a Ph.D. in animal behavior at Cambridge. Now a U.S. citizen, he came to this country to join the faculty of the University of California at Berkeley, where he directed a program of studies in animal behavior. His research concentrates on the role of bird, ape, and monkey sounds in communication. He is a member of the National Academy of Sciences and co-author of *Mechanisms of Animal Behavior*. Suggested reading: *Cooperation in Animals* by W. C. Allee; *Genetics and the Social Behavior of the Dog* by J. P. Scott and J. L. Fuller.

THE WORK OF BEING A BEE

Aubrey Manning, professor of animal behavior at the University of Edinburgh, received his B.Sc. at the University of London. At Oxford his doctoral thesis was largely concerned with the role played by the honey-guide patterns on flowers and the foraging habits of bumblebees. He has published studies on the evolution of behavior in insects and is the author of *An Introduction to Animal Behaviour*. Suggested reading: *The World of the Honeybee* by C. G. Butler; *The Dance Language and Orientation of Bees* by Karl von Frisch.

THE SOCIABLE KINGDOM

George B. Schaller, adjunct associate professor at Rockefeller University and director of the Center for Field Biology and Conservation of the New York Zoological Society, received his Ph.D. in zoology at the University of Wisconsin. Animals he has studied include gorillas in Africa, orangutans in Sarawak, and

tigers in India. He is the author of several books, including *The Deer and the Tiger, The Year of the Gorilla,* and *The Serengeti Lion,* and of articles in the April 1969 and November 1971 GEOGRAPHIC. Suggested reading: *The Emergence of Man* by John E. Pfeiffer.

SIGHTS AND SOUNDS

Richard D. Alexander, professor of zoology and curator of insects at the University of Michigan, received his Ph.D. from Ohio State University. He has taught evolutionary ecology and written on communication. In 1971 he was awarded the Daniel Giraud Elliot Medal of the National Academy of Sciences for his publications on speciation and life cycle origins in crickets. Suggested reading: *Animal Communication,* edited by Thomas Sebeok; *Animal Sounds and Communication,* edited by W. E. Lanyon and W. N. Tavolga.

TUNING IN ON THE BAT

Jack W. Bradbury, assistant professor of biology at the University of California at San Diego, received his Ph.D. at Rockefeller University. Field work has taken him to Trinidad and Costa Rica to study neotropical bats. As an assistant professor at Cornell University he continued his work by experimenting with captive bat colonies. A long-term project led him to Gabon to study the social behavior of African fruit bats. Suggested reading: *Listening in the Dark* by Donald R. Griffin; *Bats* by G. M. Allen.

CONSIDER THE ANT

Caryl P. Haskins, educator and scientist, received his Ph.B from Yale and his Ph.D. from Harvard. Founder of the Haskins Laboratories, from 1956 to 1971 he was president of the Carnegie Institution of Washington. A trustee of the National Geographic Society and the Center for Advanced Study in the Behavioral Sciences and a regent of the Smithsonian Institution, he has had a lifelong interest in social insects, especially ants. His books include *Of Ants and Men* and *Of Societies and Men.* Suggested reading: *An Introduction to the Behavior of Ants* by J. H. Sudd; *Ants* by William Morton Wheeler; *The Insect Societies* by Edward O. Wilson.

THE SONG OF THE WHALE

Roger Payne, zoologist at the New York Zoological Society's Center for Field Biology and Conservation and assistant professor at Rockefeller University, began his research career at Harvard. He received his Ph.D. at Cornell. He wrote about his studies of wild whales in the October 1972 and March 1976 GEOGRAPHIC. Suggested reading: *The Whale,* edited by L. Harrison Matthews; *The Stocks of Whales* by N. A. Mackintosh.

THE STRATEGY OF THE NICHE

Gordon H. Orians, professor of zoology at the University of Washington, spent a year as a Fulbright Fellow at Oxford before obtaining his Ph.D. from the University of California at Berkeley. He has studied the ecology of vertebrate social systems, bird community structure, and plant-herbivore interactions. He is the author of a biology textbook, *The Study of Life,* and numerous papers in scientific journals. Suggested reading: *The Biology of Populations* by Robert H. MacArthur and Joseph H. Connell.

THE ELEPHANT: LIFE AT THE TOP

John F. Eisenberg, research professor of zoology at the University of Maryland and resident scientist at the National Zoological Park, received his Ph.D. from the University of California at Berkeley. He has studied mammal behavior in the Old and New World tropics and is co-author of *The Tenrecs: A Study in Mammalian Behavior and Evolution.* Suggested reading: *The Natural History of the African Elephant* by Sylvia K. Sikes; *Some Extinct Elephants, Their Relatives and the Two Living Species* by P. E. P. Deraniyagala.

LIVING WITH MOUNTAIN GORILLAS

Naturalist Dian Fossey of California completed two years of pre-veterinary medicine before obtaining a B.Sc. She went to Africa and, on the recommendation of Dr. Louis S. B. Leakey, gained support from the National Geographic Society to carry out a study of mountain gorillas. She is working toward her doctorate in England at Cambridge, while continuing her field studies. She has published articles in the January 1970 and October 1971 GEOGRAPHIC. Suggested reading: *A Handbook of Living Primates* by J. R. and P. H. Napier; *The Mountain Gorilla* by George B. Schaller.

TERRITORY'S INVISIBLE WALLS

Richard D. Estes, professor of zoology at San Diego State University, received his B.A. in sociology at Harvard and his Ph.D. in vertebrate zoology at Cornell. In addition to his field studies of wildebeests in Africa he has made a wildlife survey of Burma and studied the sable antelope and African wild dog. He has published many articles in journals and magazines and his "Behavioral Study of East African Ungulates" appeared in the *National Geographic Society Research Reports, 1964 Projects.*

WARRING CLANS OF THE HYENA

Hans Kruuk, a senior research officer at Oxford, was born and educated in Holland. After four years of animal behavior study at Oxford, he received his Ph.D. at the University of Utrecht. From 1964 through 1971 he lived in Serengeti National Park, where he was joint founder and deputy director of the Serengeti Research Institute. He is the author of a book, *The Spotted Hyena,* and an article in the July 1968 GEOGRAPHIC. Suggested reading: *Animals of East Africa* by Louis S. B. Leakey, published by the National Geographic Society.

THE STORK: A TASTE FOR SURVIVAL

M. Philip Kahl, research scientist, received his Ph.D. in zoology from the University of Georgia. After conducting research on the stork in Florida, he extended his studies to Africa, Southeast Asia, Poland, and Argentina. He has written on the behavior of storks and is the author of an article on African flamingos in the February 1970 GEOGRAPHIC. Suggested reading: *The Life of the White Stork* by F. Haverschmidt; *The Natural Regulation of Animal Numbers* by D. Lack; *Behavioral Aspects of Ecology* by P. H. Klopfer.

THE MYSTERIES OF MIGRATION

Stephen T. Emlen, associate professor of animal behavior at Cornell University, obtained his Ph.D. in zoology at the University of Michigan. His projects include research in the mechanism of animal orientation and navigation, the ecological correlates of acoustical communication systems in birds, and the adaptive significance and strategies of social behavior in birds and mammals. His papers appear in many scientific journals. Suggested reading: *The Migrations of Birds* by Jean Dorst; *Bird Migration* by Donald R. Griffin.

THE RITES OF SPRING

John Hurrell Crook, reader in ethology at the University of Bristol, took his Ph.D. at Cambridge, England. An early interest in weaverbirds and their social organization in relation to ecology led to comparative studies of the ecological determinants of societies in birds and mammals in general. He also conducted research on primates in Africa. In 1970 he edited *Social Behaviour in Birds and Mammals.* Suggested readings in that book include the chapters "Introduction — Social Behaviour and Ethology," "The Socio-ecology of Primates," and "Gender Role in the Social System of Quelea."

COURTSHIP IN A WATERY REALM

Wolfgang Wickler, professor of zoology at the University of Munich, was born and educated in Germany and received his Ph.D. at the University of Münster. His research in the evolution and adaptive significance of behavior patterns — especially the mechanisms of pair maintenance — is carried on at the Max Planck Institute for Behavioral Physiology. His books include *Mimicry in Plants and Animals* and *Breeding Aquarium Fish.* Suggested reading: *Social Behavior from Fish to Man,* edited by William Etkin.

THE WAY OF THE BISON

Dale F. Lott, chairman of the department of wildlife and fisheries biology and professor of psychology at the University of California at Davis, received his Ph.D. from the University of Washington and was a post-doctoral Fellow at Rutgers Institute of Animal Behavior. His career, which started in the laboratory, has shifted to field studies of relationships between social organizations and social behavior and ecology. Suggested reading: *Evolution of Behavior* by Jerram Brown; *Ethology of Mammals* by R. F. Ewer; *Courtship* by M. Bastock.

411

IN A PENGUIN COLONY

Richard L. Penney received his Ph.D. at the University of Wisconsin. He taught and conducted research in pathobiology at Johns Hopkins University. He was assistant director, Institute for Research in Animal Behavior, New York Zoological Society and Rockefeller University. He was National Science Foundation (Polar Programs) representative to New Zealand, 1970-71. He is the author of *The Penguins Are Coming!*

LEARNING HOW ANIMALS LEARN

Arthur J. Riopelle, professor of psychology at Louisiana State University, received his Ph.D. from the University of Wisconsin. He has served as director of the Delta Regional Primate Research Center of Tulane University. He is editor of the book *Animal Problem Solving* and author of articles on Snowflake, the world's first known white gorilla, in the March 1967 and October 1970 GEOGRAPHIC. Suggested reading: *Animal Behavior* by Keller Breland and Marian Breland.

LEARNING TO LIVE

William A. Mason, professor of psychology at the University of California at Davis and senior research psychologist at the National Center for Primate Biology, received his Ph.D. from Stanford University. His research studies include social behavior in young chimpanzees. His articles have appeared in books and professional journals. Suggested reading: *Social Behaviour in Animals* by N. Tinbergen; *Early Experience and Behavior,* edited by G. Newton and S. Levine.

QUEST FOR THE ROOTS OF SOCIETY

Irven DeVore, professor of anthropology at Harvard University, received his Ph.D. from the University of Chicago. His field work includes research on baboons and orangutans, and continuing study of the Bushmen of the Kalahari Desert. He edited *Primate Behavior* and is co-author of *The Primates* and co-editor of *Man the Hunter.* Suggested reading: *In the Shadow of Man* and *My Friends the Wild Chimpanzees* by Jane van Lawick-Goodall; *Primates,* edited by Phyllis C. Jay.

Acknowledgments

The editors appreciate assistance by Helmut K. Buechner, J. F. Gates Clarke, Robert H. Gibbs, H. W. Setzer, George E. Watson, and Alexander Wetmore, Smithsonian Institution; Lincoln Pierson Brower, Amherst College; Masakazu Konishi, Princeton University; Burney J. Le Boeuf, University of California; Lorus J. and Margery J. Milne, University of New Hampshire; Craig Phillips, National Aquarium; David and Ann Premack, University of California; Edward S. Ross, California Academy of Sciences; Thomas W. Schoener, Harvard University; Niko Tinbergen, University of Oxford.

Animals in this book

In this selective taxonomic, or classification, list, the first Latin name given is the genus, the second is the species, and the third is the race, or subspecies. Where no one species of animal is meant, the genus, family, or order alone is given. Where more than one name exists, the choice expresses the consensus of authorities.

COMMON NAME	SCIENTIFIC NAME
Mammals	
Baboon, gelada	*Theropithecus gelada*
sacred	*Papio hamadryas*
savanna	*Papio cynocephalus*
Bat, European horseshoe	*Rhinolophus ferrum-equinum*
fishing	*Noctilio leporinus*
flying fox (fruit)	genus *Pteropus*
greater spearnosed	*Phyllostomus hastatus*
hammerheaded	*Hypsignathus monstrosus*
lesser spearnosed	*Phyllostomus discolor*
Pel's	*Taphozous peli*
vampire	*Desmodus rotundus*
whitelined	*Saccopteryx bilineata*
Bear, polar	*Thalarctos maritimus*
sloth	*Melursus ursinus*
Bison	*Bison bison*
Buffalo, Cape	*Syncerus caffer*
Bushbaby, lesser	*Galago senegalensis moholi*
Caribou	*Rangifer tarandus*
Cat, domestic	*Felis catus*
Cheetah	*Acinonyx jubatus*
Chimpanzee	*Pan troglodytes*
Coyote	*Canis latrans*
Dog, African wild	*Lycaon pictus*
domestic	*Canis familiaris*
Elephant, African	*Loxodonta africana*
Asian	*Elephas maximus*
Gazelle, Thomson's	*Gazella thomsoni*
Giraffe	*Giraffa camelopardalis*
Gorilla, lowland	*Gorilla gorilla gorilla*
mountain	*Gorilla gorilla beringei*
Hippopotamus	*Hippopotamus amphibius*
Horse	*Equus caballus*
Hyena, spotted	*Crocuta crocuta*
Impala	*Aepyceros melampus*
Jackal, black-backed	*Canis mesomelas*
Kob, Uganda	*Kobus kob thomasi*
Lion	*Panthera leo*
Macaque, Japanese	*Macaca fuscata*
rhesus	*Macaca mulatta*
Mongoose	genus *Herpestes*
Moose	*Alces alces*
Musk ox	*Ovibos moschatus*
Opossum	*Didelphis marsupialis*
Otter, sea	*Enhydra lutris*
Pangolin	genus *Manis*
Prairie dog, black-tailed	*Cynomys ludovicianus*
Rabbit, European	*Oryctolagus cuniculus*
Raccoon	*Procyon lotor*
Rhinoceros, black	*Diceros bicornis*
white	*Ceratotherium simus*
Sable antelope	*Hippotragus niger*
Sea lion, Steller	*Eumetopias jubata*
Seal, leopard	*Hydrurga leptonyx*
northern elephant	*Mirounga angustirostris*
southern elephant	*Mirounga leonina*
Sheep	genus *Ovis*
Tiger	*Panthera tigris*
Topi	*Damaliscus lunatus*
Whale, beluga (white)	*Delphinapterus leucas*
blue	*Balaenoptera musculus*
bottle-nosed, northern	*Hyperoodon ampullatus*
bottle-nosed, southern	*Hyperoodon planifrons*
bowhead (Greenland right)	*Balaena mysticetus*
false killer	*Pseudorca crassidens*
finback	*Balaenoptera physalus*
goose-beaked	*Ziphius cavirostris*
gray	*Eschrichtius robustus*
humpback	*Megaptera novaeangliae*
killer	*Orcinus orca*
little piked	*Balaenoptera acutorostrata*
narwhal	*Monodon monoceros*
pilot	*Globicephala melaena*
pygmy right	*Caperea marginata*

pygmy sperm	*Kogia breviceps*
right	*Eubalaena glacialis*
sei	*Balaenoptera borealis*
sperm	*Physeter catodon*
dusky dolphin	*Lagenorhynchus obscurus*
Wildebeest	*Connochaetes taurinus*
Wisent	*Bison bonasus*
Wolf	*Canis lupus*
Zebra, Burchell's	*Equus burchelli*

Birds

Blackbird, Brewer's	*Euphagus cyanocephalus*
red-winged	*Agelaius phoeniceus*
tricolored	*Agelaius tricolor*
yellow-headed	*Xanthocephalus xanthocephalus*
Booby, brown	*Sula leucogaster*
Bowerbird, satin	*Ptilonorhynchus violaceus*
Bunting, indigo	*Passerina cyanea*
Chaffinch, European	*Fringilla coelebs*
Chicken	*Gallus gallus*
Crane, sandhill	*Grus canadensis*
Duck	family Anatidae
Eagle, bald	*Haliaeetus leucocephalus*
Finch, ground	genus *Geospiza*
woodpecker	*Camarhynchus pallidus*
Flamingo, American (greater)	*Phoenicopterus ruber*
lesser	*Phoenicopterus minor*
Fowl, jungle	*Gallus lafayettii*
mallee	*Leipoa ocellata*
Frigatebird	family Fregatidae
Grebe, great crested	*Podiceps cristatus*
Greenfinch	*Chloris chloris*
Gull, black-headed	*Larus ridibundus*
laughing	*Larus atricilla*
Hawk, red-tailed	*Buteo jamaicensis*
Heron, black-headed	*Ardea melanocephala*
Hornbill, red-billed	*Tockus erythrorhynchus*
Jay, blue	*Cyanocitta cristata*
Killdeer	*Charadrius vociferus*
Kittiwake	genus *Rissa*
Manakin	*Manacus manacus*
Owl, barn	*Tyto alba*
screech	*Otus asio*
Oxpecker, red-billed	*Buphagus erythrorhychus*
Penguin, Adelie	*Pygoscelis adeliae*
emperor	*Aptenodytes forsteri*
Pigeon	family Columbidae
Plover, golden	*Pluvialis dominica*
Rook	*Corvus frugilegus*
Skimmer, black	*Rynchops nigra*
Skua	*Catharacta skua*
Snakebird	*Anhinga rufa*
Sparrow, house	*Passer domesticus*
white-crowned	*Zonotrichia leucophrys*
Stork, Abdim's	*Sphenorhynchus abdimii*
African openbill	*Anastomus lamelligerus*
American wood	*Mycteria americana*
Asian openbill	*Anastomus oscitans*
black	*Ciconia nigra*
maguari	*Euxenura galeata*
marabou	*Leptoptilos crumeniferus*
painted	*Ibis leucocephalus*
saddlebill	*Ephippiorhynchus senegalensis*
white	*Ciconia ciconia*
woolly-necked	*Dissoura episcopus*
yellowbilled	*Ibis ibis*
Tern, royal	*Thalasseus maximus*
sooty	*Sterna fuscata*
Thrush, Swainson's	*Catharus ustulata*
Tit, blue	*Parus caeruleus*
Vulture, Egyptian	*Neophron percnopterus*
hooded	*Necrosyrtes monachus*
Warbler, bay-breasted	*Dendroica castanea*
Blackburnian	*Dendroica fusca*
blackpoll	*Dendroica striata*
black-throated green	*Dendroica virens*
Cape May	*Dendroica tigrina*
magnolia	*Dendroica magnolia*
myrtle	*Dendroica coronata*
Baya weaver	*Ploceus philippinus*
Quelea weaver	*Quelea quelea*
Village weaver	*Ploceus cucullatus*

Reptiles

Boa, emerald tree	*Corallus caninus*
Cobra	*Naja naja*
Crocodile, Nile	*Crocodylus niloticus*

Lizards	
anole	*Anolis evermanni*
anole	*Anolis gundlachi*
anole	*Anolis krugi*
anole	*Anolis occultus*
anole	*Anolis stratulus*
ground goanna (monitor)	*Varanus goldii*
iguana, land	*Conolophus subcristatus*
marine	*Amblyrhynchus cristatus*
Parson's chameleon	*Chamaeleo parsoni*
Turtle, green	*Chelonia mydas*

Amphibians

Frog	family Ranidae
cricket frog	*Acris gryllus*
Salamander	family Ambystomidae
Toad	family Bufonidae

Fishes

Cichlids	
African jewelfish	*Hemichromis bimaculatus*
African mouthbrooder	*Haplochromis burtoni*
Congo river cichlid	*Lamprologus congolensis*
discus fish	*Symphysodon discus*
Egyptian mouthbrooder	*Haplochromis strigigena*
firemouth	*Cichlasoma meeki*
orange chromide	*Etroplus maculatus*
spotted mouthbrooder	*Tropheus moorei*
tilapia	genus *Tilapia*
veiled angel	*Pterophyllum eimekei*
Eel, European	genus *Anguilla*
moray	family Muraenidae
Salmon, sockeye	*Oncorhynchus nerka*
Siamese fighting fish	*Betta splendens*
Snapper, gray	*Lutjanus griseus*
Stargazer, southern	*Astroscopus y-graecum*
Stickleback, three-spined	*Gasterosteus aculeatus*
Swordtail	genus *Xiphophorus*

Insects and other invertebrates

Ant, aphid-tending	*Prenolepis imparis*
army, New World	genus *Eciton*
bulldog	genus *Myrmecia*
driver, African	genus *Dorylus*
fungus-growing	genus *Atta*
fungus-growing	genus *Trachymyrmex*
harvester	*Pogonomyrmex badius*
honey	genus *Myrmecocystus*
slave-making	*Formica pergandei*
slave-making (wood ant)	*Formica rufa*
slave-making	*Formica sanguinea*
slave-making	genus *Polyergus*
thief, African	*Carebara vidua*
weaver	genus *Oecophylla*
Aphid	order Homoptera
Barnacle	order Cirripedia
Bees	
bumblebee	genus *Bombus*
honeybee	*Apis mellifera*
solitary bee	genus *Megachile*
Beetle, deathwatch	family Anobiidae
dung	family Scarabaeidae
Butterfly, monarch	*Danaus plexippus*
spicebush swallowtail	*Papilio troilus*
Cicada	family Cicadidae
Coral	order Zoantharia
Cricket, field	genus *Gryllus*
pine-tree	*Oecanthus pini*
snowy tree	*Oecanthus fultoni*
Firefly, Southeast Asian	*Pteroptyx malaccae*
Katydid, false	subfamily Phanepterinae
true	*Pterophylla camellifolia*
Mantis, Malayan	*Hymenopus coronatus*
Moth, cabbage looper	*Trichoplusia ni*
gypsy	*Porthetia dispar*
saddle-back	*Sibine stimulea*
Octopus	*Octopus briareus*
Portuguese man-of-war	*Physalia physalis*
Robber fly	family Asilidae
Sawfly	genus *Neodiprion*
Scallop, bay	*Pecten irradians*
Shrimp, cleaner	suborder Natantia
Squid	*Loligo pealei*
Starfish, ocher	*Pisaster ochraceus*
Stinkbug	family Pentatomidae
Termite	order Isoptera
Worm, palolo	*Eunice viridis*

413

INDEX
Text references appear in <u>lightface type,</u> illustrations and illustrated text in **boldface.**

Composition by National Geographic's Photographic Services, Carl M. Shrader, Director, Lawrence F. Ludwig, Assistant Director. Color separations by Beck Engraving Company, Philadelphia, Pa., Chanticleer Company, Inc., New York, Colorgraphics, Inc., Beltsville, Md., Graphic Color Plate, Inc., Stamford, Conn., The Lanman Company, Alexandria, Va., Lebanon Valley Offset, Inc., Cleona, Pa., Progressive Color Corporation, Rockville, Md., Stevenson Photo Color Company, Cincinnati, Ohio. Printed and bound by Kingsport Press, Kingsport, Tenn. Paper by Westvaco Corporation, New York.

DRAWING BY GEORGE FOUNDS